Solid-State
Circuit Analysis

Solid-State Circuit Analysis

VESTER ROBINSON

RESTON PUBLISHING COMPANY, INC.
A Prentice Hall Company
Reston, Virginia   22090

Library of Congress Cataloging in Publication Data

Robinson, Vester.
   Solid state circuit analysis.
   xii, 372 p.
   Bibliography: p. 328–329.
   Includes index.
      1. Electronics.   2. Semiconductors.   3. Transistors.   I. Title.
TK7815.R59   621.3815   75-1495
ISBN 0-87909-769-8

© 1975 by Reston Publishing Company, Inc.
A Prentice-Hall Company
Reston, Virginia 22090

All rights reserved. No part of this book may be reproduced in any way, or by any means, without permission in writing from the publisher.

1 3 5 7 9 10 8 6 4 2

PRINTED IN THE UNITED STATES OF AMERICA

To a great little woman, my wife, Florence Belle

Contents

Preface

Chapter One
Solid State Devices   1

Basic Theory 1,   Introduction to Diodes 4,   Common Diodes 5,
Tunnel Diodes 12,   Zener Diodes 15,   Silicon Controlled
Rectifiers 16,   Photodiodes 18,   Introduction to Transistors 19,
Bipolar Transistors 19,   Field Effect Transistors 22,   *Review
Exercises 27*

Chapter Two
Energy Conversion Circuits   30

Power Transformers 30,   Power Supply Rectifiers 38,   Rectifier
Circuits 39,   Power Supply Filters 44,   Voltage Multipliers 47,
Voltage Dividers 49,   Voltage Regulators 53,   Converters 59,
Review Exercises 62

Chapter Three
Principles of Amplification   65

Basic Amplifiers 65,   Common Base 68,   Common Emitter 74,
Common Collector 80,   Bias Stabilization 84,   *Review Exercises 91*

Chapter Four
Small Signal Amplifiers   95

Dynamic Characteristics 95,   Open Circuit Parameters 97, Short
Circuit Parameters 100,   Hybrid Parameters 102,   Parameter
Interrelationship 109,   Applying Parameters 112,   *Review
Exercises 113*

Chapter Five
Large Signal Amplifiers 115
Load Lines 115, Dynamic Transfer 124, Power Dissipation 128, Frequency Problems 131, Other Amplifier Circuits 133, *Review Exercises 140*

Chapter Six
Audio Amplifiers 143
Current and Voltage Distribution 143, Preamplifiers 149, Interstage Coupling 154, Circuit Controls 159, Audio Drivers 161, Power Amplifiers 164, *Review Exercises 171*

Chapter Seven
Tuned Amplifiers 174
Interstage Coupling 174, Feedback Arrangements 185, Automatic Gain Control 189, *Review Exercises 193*

Chapter Eight
Wide Band Amplifiers 196
Signal Frequencies 196, Types of Coupling 198, Frequency Response 201, Low Frequency Response 202, High Frequency Response 207, *Summary 215*, *Review Exercises 215*

Chapter Nine
Nonlinear Functions 218
Nonlinear Devices 218, Function Generators 223, Logarithmic Amplifiers 229, *Review Exercises 234*

Chapter Ten
Sinusoidal Oscillators 237
Principles of Oscillation 237, Resonant Feedback Oscillators 241, Crystal Controlled Oscillators 250, Nonresonant Feedback Oscillators 253, *Review Exercises 257*

Chapter Eleven
Nonsinusoidal Oscillators 260
Relaxation Oscillators 260, Synchronized Oscillators 269, *Review Exercises 278*

Chapter Twelve
Modulators 281

Microphones 281, Amplitude Modulation Principles 285,
Amplifier Modulators 288, Oscillator Modulators 292,
Frequency Modulation Principles 295, Frequency Modulators 300,
Review Exercises 302

Chapter Thirteen
Demodulators 305

Receiver Functions 305, Mixers and Converters 309,
AM Demodulators 314, FM Demodulators 318, Audio
Reproduction 322, *Review Exercises 325*

Bibliography 328

Answers to Review Exercises 330

Index 363

Preface

This book is intended for use as a fundamental text in the analysis of solid state electronic circuits. The prerequisites to its use are high school mathematics and an introductory course in electricity and electronics. My own programmed texts on these subjects or similar ones will provide that necessary background.

My primary aim in this book is to prepare the undergraduate for advanced courses in electronic maintenance and circuit design. The approach is eminently practical. Most of the basic principles of electronic circuits are learned by analyzing the circuits which are commonly found in a great variety of electronic equipment.

The first chapter is an overview of solid state devices. Basic theory, structure, and characteristics are covered in nontechnical terms. Several types of solid state diodes are discussed along with bipolar and field effect transistors. The remaining chapters concentrate on the analysis of practical circuits which use solid state devices as circuit components. Mathematics are used when an equation or a calculation is essential to the understanding of a circuit; however, the emphasis is on visualization of relative values rather than on rigid calculation of precise values.

Many people consider electronics a complex subject, consequently many textbooks are written at a level beyond the comprehension of the beginning student. These two facts add to the burdens of the student and the teacher. Both in learning and in teaching I have become familiar with the extremes of these problems. In this text, I try to combine my own knowledge and research with nearly 30 years of teaching experience to fill that middle ground between the too simple and the too complex approach to electronic circuits, especially solid state circuits. I mean this book to be easy to understand so that the confidence and knowledge gained through its use will enable the student to master more advanced courses with considerably less effort than was previously required.

*   *   *

I extend my appreciation to many people for their assistance in compiling this book: to the experts in many electronic firms who have been extremely generous about supplying technical data; to Matthew I. Fox, the president of Reston Publishing Company, and to his editors and artist who have been regularly helpful with their professional guidance and assistance; and to David Summers and Herbert W. Jackson who reviewed the first draft of this manuscript. These two, in particular used their considerable technical knowledge and careful analysis to make many valuable suggestions to the author, thus greatly improving the book.

VESTER ROBINSON

Chapter One

Solid State Devices

All the circuits in this book have one factor in common; they all use at least one solid state device as a circuit component. The devices include common diodes, Zener diodes, silicon controlled rectifiers, tunnel diodes, bipolar transistors, unijunction transistors, and field effect transistors of the metal oxide type. In this chapter we shall briefly review the basic solid state theory and discuss the functional characteristics of these specific components.

BASIC THEORY

There are three categories of materials: conductors, semiconductors, and insulators. All of our materials fit into one of these three categories. On one end, we have the metals with many free electrons. We call these conductors because they form an easy path for current. At the other extreme, we have materials such as rubber, Bakelite, and polystyrene, which have very few free electrons. These materials offer a strong opposition to current, and we call them insulators. In the middle we have a wide variety of materials that are neither good conductors nor good insulators. We call them semiconductors. The characteristics of these semiconductors have enabled the development of the whole area of solid state devices.

Resistivity The resistance of a material is directly proportional to its quality as an insulator. The good insulators have an extremely high resistance to the flow of electrons. The good conductors, on the other hand, offer very low resistance to this electron flow. The resistivity of semiconductors fall in the middle region; they offer more resistance than the good conductors but less than the good insulators. If we plot the resistivity of some sample materials on a graph, it will be similar to that in Fig. 1-1.

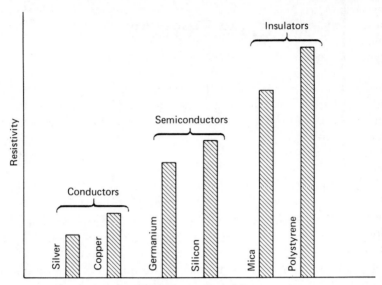

FIGURE 1-1. Resistivity

Semiconductor Crystals Most solid materials exhibit one of two types of structure: cell or crystal. Biological structures such as wood, leaves, bones, and tissue are examples of materials that are composed of cells. Most other materials have a crystalline structure. We do not normally think of stones and metals as being crystals, but the crystalline pattern is present.

The atoms in a crystal are arranged in a specific pattern. We find that each atom is closely related to a number of adjacent equidistant atoms. The specific arrangement depends on size and number of atoms and on the electrical forces between the atoms. The arrangement and type of atoms determine the physical, electrical, optical, and mechanical characteristics of the crystal. All materials in the semiconductor and conductor classes are crystalline substances. The conductors are polycrystalline materials, and the semiconductors are single-crystal materials. Some of these semiconductors are suitable for constructing solid state devices for use in electronic circuits. The two materials most frequently used are silicon and germanium.

Current Carriers In the pure state, both silicon and germanium are very stable elements. The atoms of each have four electrons in the outer (valence) band, and they share four other electrons with adjacent atoms. This covalent bonding gives each atom the effect of having a full (eight electrons) outer band. Materials with eight electrons in the outer band are good insulators; therefore, silicon and germanium are insulators when in the pure state. If we would have these two materials carry current, we must do something to alter the covalent bond of the atoms in the crystal. We do this by doping the crystalline melt with small quantities of other materials.

Some materials, indium for instance, have only three valence electrons in each atom. We call these materials trivalent materials. Other materials have a pentavalent structure: five electrons in the valence band of each atom. Arsenic is an example of a pentavalent element. We can dope a melt of pure silicon or germanium with arsenic or indium to create the semiconductor characteristics that we desire.

When we dope a melt with a small quantity of indium (or other trivalent material), all crystals grown from this melt will have a shortage of electrons in the valence band of a portion of the atoms. Since a hole is equivalent to a missing electron, we could say that these crystals have a surplus of holes. Each hole is a small positive charge that will attract and capture any free electron in its vicinity. We call this acceptor material and normally designate it as P-type material. The current carriers in P-type material are the holes.

When we dope a melt of pure silicon or germanium with a small quantity of arsenic (or other pentavalent material), all crystals grown from this melt will have a surplus of electrons in the valence band of a portion of the atoms. We call this donor material and normally designate it as N-type material. The current carriers in N-type material are the electrons.

Electrons will flow through both P- and N-type materials when a voltage is placed across them, as illustrated in Fig. 1-2.

The carriers (holes) in the P-type material are represented by plus signs. These positive holes move from the material to the negative battery terminal and from the positive terminal back into the material. The electrons, which compose our current, move in the opposite direction to the movement of the holes. The carriers (electrons) in the N-type material are represented by minus signs. Now both carriers and current are composed of electrons with direction of movement as indicated in the illustration.

Electron Current versus Conventional Current Conventionally the electrical and electronics industry has considered current to move from positive to negative. More recently, many writers have adopted the idea that the movement of electrons is identical to current. In fact, the definition of an ampere of current is the movement of 10^{18} electrons past a point per second. The definition of electron current then is simply the flow of electrons.

4 Solid State Devices

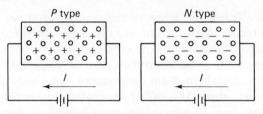

FIGURE 1-2. Movement of Electrons

Electron current is used throughout this text. The single word, "current," will mean "flow of electrons." If you are more familiar with conventional current, you need not change your concepts. Just remember that current in this book is electron flow, and conventional current moves in the opposite direction.

Joining Materials Diodes and transistors are formed by joining P- and N-type materials. The junction between P- and N-type materials is formed by one of several chemical processes. The result *is not* two layers placed together like a sandwich. The process forms one compact crystal, a single unit, with P-type material on one side and N-type material on the other. The area to either side of the junction will be depleted of both positive and negative carriers, and we call this the depletion area. Outside the depletion area, we have N-type material in one direction and P-type material in the other direction. With a little encouragement, electrons will move easily across the junction from the N material to the P material. But there is a very stubborn opposition to movement of electrons from the P material across the junction to the N material. This is the rectifying characteristic of our solid state diode. We now have a device which passes current (electrons) in only one direction: from N to P. This is a rectifying diode.

INTRODUCTION TO DIODES

The principal task of a diode is rectification. The rectifying ability of solid state materials was recognized even before the invention of electron tubes, but discovery of the tubes diverted attention away from solid state devices. This diversion delayed the invention of the transistor for many years. Today we have several other types of diodes in addition to the common rectifying diode. For example, we have Zener diodes, tunnel diodes, silicon controlled rectifiers, photodiodes, and others. Let us examine some of these diodes in greater detail.

COMMON DIODES

For lack of a better term, we shall refer to the familiar rectifier diodes as *common* diodes. We find these diodes used in power supply rectifiers, frequency discriminators, clampers, gating circuits, and many other everyday applications in electronic circuits.

Inducing Current Through the Junction When we attach a battery across a PN junction as shown in Fig. 1-3, current will exist through the junction from the N- to the P-type material. Electrons are pushed from the negative terminal of the battery into the N-type material. This forces electrons to cross the junction and fill holes in the P-type material. The positive side of the battery attracts electrons from the P-type material and keeps the supply of holes replenished.

There is very little opposition to current in this direction. A very small voltage can create a rather large current through the junction. The battery polarity is aiding the current carriers, and we say that the diode junction is forward biased.

If we reverse the polarity of the battery, as shown in Fig. 1-4, current will be almost completely blocked. In this condition both the holes and the electrons are drawn away from the junction. The negative terminal of the battery is a strong attraction for the holes, while the positive terminal offers the same strong attraction for the electrons. As the carriers are pulled away from the junction, practically the area to either side of the junction is left void of both types of current carriers. The semiconductor in the vicinity of the junction has become an insulator and very effectively blocks current. Since the battery polarity is now opposing the carriers, we call this a reverse biased junction.

Of course, we always have a few random carriers near the junction, and they allow a very minute current from the P- to the N-type material. This current is on the order of a few microamperes,

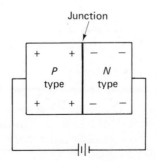

FIGURE 1-3. Obtaining a Flow of Electrons

6 Solid State Devices

and we call it leakage current. The leakage current is so small that we properly ignore it in most practical applications.

Forward Biased Junction The primary purpose of P-type material is to generate holes and conduct these holes to the PN junction. The fact that electrons must flow through the material to fill these holes is of secondary importance. These holes are our positive current carriers. Conversely, the primary purpose of N-type material is to generate excess electrons and conduct these electrons to the PN junction. The fact that holes must flow through the N-type material is of secondary importance. The electrons are our negative current carriers.

When we place a voltage across the junction with the polarity aiding both positive and negative current carriers, as shown in Fig. 1–5, the junction is forward-biased. Notice that the forward bias condition is obtained by connecting the positive battery terminal to the P-type material and the negative battery terminal to the N-type material. The negative terminal repels electrons through the N-type material toward the PN junction. The positive terminal repels holes through the P-type material toward the PN junction. By the same token, the negative terminal attracts holes through the crystal while

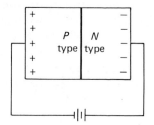

FIGURE 1–4. Blocking the Flow of Electrons

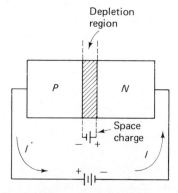

FIGURE 1–5. Forward Bias

the positive terminal attracts electrons. This action is a recombining process which keeps electrons flowing from the negative battery terminal, through the crystal, and back to the positive battery terminal.

The depletion region exhibits a characteristic similar to a small battery with negative toward the P-type material and positive toward the N-type material. We frequently refer to this characteristic as either *space charge* or *barrier potential*. The potential is directly proportional to the thickness of the depletion region. Placing an external battery with negative to the N-type material and positive to the P-type material (forward bias) has the effect of reducing the thickness of the depletion region and thereby reducing the barrier potential.

A very slight external voltage causes a considerable current in this direction. If we increase the external voltage, the space charge (barrier potential) will completely disappear. When this happens, the only barrier to current is the resistance of the crystal. Figure 1–6 is a characteristic curve for a forward-biased PN junction of both germanium and silicon.

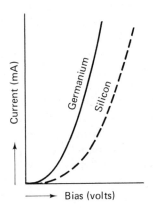

FIGURE 1-6. Forward-biased PN Junctions

The difference in current levels is accounted for by the fact that silicon has a higher resistivity than germanium. Notice that when our forward bias reaches a certain potential, a very small increase in bias causes a very high increase in current. This is the point where the depletion region is very thin and the barrier potential is almost cancelled out. If we increase the bias until the barrier potential is completely nullified, our diode will be destroyed. The excessive current will generate a high temperature, and this heat will damage the junction. This provides current paths through the junction and eliminates the rectifying qualities of the diode.

Reverse-biased Junction When we connect an external battery with the negative terminal to the P-type material and the positive terminal to the N-type material, the junction is reverse-biased. This condition is illustrated in Fig. 1-7.

Our reverse bias condition opposes both the positive and negative current carriers. The negative terminal of the external battery attracts the hole carriers in the P-type material away from the junction. The positive terminal of the battery attracts the electron carriers in the N-type material away from the junction. The result is a thicker depletion region, which is equivalent to a higher voltage space charge (barrier potential).

Within the working limits of the diode, the barrier potential is equal to the bias potential. Since these potentials are series opposing, they cancel each other, and the result is practically no current in our circuit. True; there is a small leakage current on the order of a few microamperes, but we can, and do, safely ignore this reverse current in most practical applications. Increasing the bias voltage simply increases the thickness of the depletion region, increases the barrier potential, and still produces a negligible quantity of current. Of course, there is a limit. If we continue increasing the reverse bias, we will reach a point where current will arc through the crystal. This is the breakdown potential. When we subject a common diode to the breakdown potential, the junction will be damaged, and the rectifying qualities of the diode will be destroyed. This high reverse potential drives the carriers with such a high velocity that they bombard and release large quantities of additional carriers from the valence bonds of the crystals. This action sets up a chain reaction, which rapidly produces a virtual avalanche of carriers. We refer to this action as avalanche breakdown. After avalanche breakdown, the only opposition to current is the resistance of the crystal. We now have a

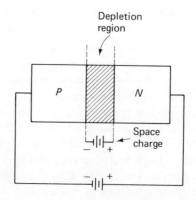

FIGURE 1-7. Reverse-biased PN Junctions

very high current in the reverse direction, which we call avalanche current. This condition is represented on the graph in Fig. 1–8.

Here we place our graph in the lower left quadrant to illustrate reverse bias and current in the reverse direction. Starting from zero volts, we can increase the reverse bias all the way to the breakdown potential with a very slow increase in reverse (leakage) current. Once our reverse bias reaches this breakdown potential, our reverse current instantly jumps to a very high level. The breakdown voltage is illustrated on the graph, and it brings the current to point X on the current curve. This is the avalanche breakdown point of voltage and current. The current beyond point X is avalanche current.

Junction Temperature The temperature of the junction has a direct effect on the electrical characteristics of the PN crystal diode. The creation of additional carriers (both holes and electrons) depends upon the breaking of electron pair bonds within the crystal. This can be done by either electrical or thermal energy.

When we subject a PN junction to high ambient temperatures or high temperatures caused by current through the junction, the diode may become unstable. If we carry this action to extremes, the diode will be destroyed. The heating effect can become regenerative to the point of destroying the junction. We call this condition *thermal runaway*. Remember that the power losses (I^2R) of the diode are dissipated in the form of heat energy. Increasing the heat energy breaks a few electron pair bonds. This action releases more heat energy while creating additional current carriers. The action is accumulative, resulting in more current and more heat. The added carriers and increased current cause more power losses. The regenerative process continues until thermal runaway causes our diode to become unstable and erratic. This is the minimal effect of thermal runaway. If it is allowed to continue, the junction will be destroyed.

Classification of Common Diodes Common rectifier diodes have a wide variety of electronic circuit applications. They are

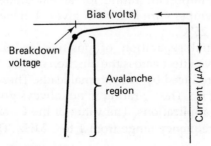

FIGURE 1–8. Reverse-biased Junctions

used for power supply rectifiers, demodulation circuits, electronic switching, signal clamping, and many other functions. We can classify our common rectifier diodes, for all these purposes, as either signal diodes or power supply diodes.

Our *signal diode* is designed to operate under small signal conditions. We find them in demodulators, switching arrangements, and other similar applications. The junctions of these diodes are extremely thin to provide reliable operation at high switching speeds. This thin junction also improves the response of the diode at high frequencies.

Our power supply diodes are expected to function under the control of direct voltage. They are designed to withstand larger amounts of power. The junctions are made large in cross-sectional area to allow large currents with minimum resistance. Provisions are also made for dissipating the additional heat that results from the increased power. Cooling fins and heat sinks are among the methods used for heat removal. When our circuit requirements are low forward voltage drops and moderately low temperatures, we would probably select a germanium diode because of its low resistance. When we have high voltages or anticipate high operating temperatures, the higher resistance silicon units are more desirable.

Ratings for Signal Diodes Manufacturers rate signal diodes in accordance with the function that they are expected to perform. When the diode is expected to function under dc control, maximum values are given. Working voltage (WVDC) is the voltage that should produce the best operating conditions. Forward current I_f is designated at a given voltage and a standard ambient temperature. The standard ambient temperature is normally specified; if not, it is understood to be 25 degrees celcius (77°F). We often find I_f ratings similar to this: 40 mA/25°C/1 volt. This rating indicates a maximum forward current of 40 mA when the ambient temperature is 25°C and the applied voltage is 1 volt. This rating in no way implies the maximum voltage, current, or temperature.

Another important rating for signal diodes is breakdown voltage B_v. The temperature is seldom given, but we understand it to be the standard, 25°C.

The spatial separation of positive and negative charges causes a capacitive effect across the junction of a diode. Other capacitive effects are produced by our external leads. These effects lead to a capacitive rating C_0. The C_0 rating is not always given, but it is important in some applications. The value of the C_0 rating is generally stipulated at a frequency range from 1 to 5 MHz. This rating affects the action when one is handling sinusoidal waves as well as the switching response characteristics. The shape of the sine wave is not altered, but the capacitive reactance reduces the amplitude of the output. The capacitive effect has a greater impact on pulse voltages. A pulse shape will be altered. The capacitive effect causes slopes on both leading and trailing edges and produces overshoot spikes as well.

Signal diodes have a recovery time rating for both leading and trailing edges. These are of prime importance when we are planning to use the diode in a switching circuit. Forward recovery time is associated with the leading edge of the pulse, while reverse recovery time applies to the trailing edge. Recovery time is closely related to capacitance, but other factors are involved.

Ratings for Power Supply Diodes Power supply rectifier diodes are designed to deliver a great deal more power than the signal diodes. For this reason, we find that their junctions are formed with a wide cross-sectional area. This provides minimum crystal resistance and permits the passage of high currents. Along with increased power handling capabilities, provision must be made for dissipating greater quantities of heat.

Peak inverse voltage (PIV) is one of the important ratings for a power supply diode. This is the maximum peak value of voltage that can be safely applied to the diode in the reverse direction. If we subject a diode to reverse voltages in excess of PIV, we should expect it to break down.

In regard to current ratings of a diode, we have two important ratings to consider: recurrency current and transient effects current. We may find a recurrent peak forward current rating which reads simply I_f 150 mA. This means that the diode is to handle a maximum peak current of 150 mA at an ambient temperature of 25°C. Sometimes the rating is for temperatures other than the 25° standard; in this case, the temperature is specified. For instance: I_f 50 A/50°C. This recurrent rating is based on the assumption that the frequency applied to the diode is 60 Hz. A diode which handles an I_f of 50 A at 60 Hz cannot possibly handle the same I_f at a frequency of 400 Hz. Surge current is a term used to describe transient effects. This is also a peak forward current rating normally designated as I_f (surge). With surges, the duration of the surge is just as important as the amplitude. For this reason, we find the pulse width specified in the surge ratings. Here is a sample rating: I_f (surge) 1 A/1 μs. It means that this diode can withstand forward surges of 1 A peaks provided the pulse width does not exceed 1 μs.

Occasional forward surges of current in a power supply

12 Solid State Devices

seldom cause damage to our diodes, even when these surges exceed the rating of the diode. Reverse voltage surges that exceed the PIV rating are more likely to destroy the diode. These reverse voltage surges frequently drive the rectifier past the breakdown point and result in diode failure.

Schematic Symbol The schematic symbol for the common rectifier diode is illustrated in Fig. 1-9. This symbol is used for both signal and power supply diodes. Since we are dealing with electron flow as current, the forward current passes through the diode against the arrow as indicated. The polarity designations, anode-cathode labels, and current direction *are not* part of the symbol.

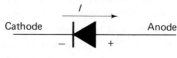

FIGURE 1-9. Diode Symbol

TUNNEL DIODES

The *tunnel diode* is so named because it operates on a principle of quantum mechanics that we call *tunneling*, or the tunnel effect. The tunnel diode is basically a PN junction with three special characteristics. It has an extremely thin barrier, a negative resistance area, and a large concentration of carriers.

Tunnel Effect The tunnel effect is easier to understand if we compare the behavior of carriers in a common diode with those in a tunnel diode. Refer to Figs. 1-10 and 1-11 for this comparison.

Graph A indicates that a certain quantity of energy must be imparted to an electron in order to push it over the potential hill. This is the energy required to overcome the space charge and move an electron across the barrier from the N- to the P-type material.

Graph B is the same type of information concerning the hole carriers. We see that a certain quantity of energy is required to move each hole over the potential hill from the P- to the N-type material.

We previously described this action of carriers as a diffusion process. This action can be illustrated by placing a drop of ink into a glass of water. As forward bias increases, the barrier becomes lower and diffusion increases. Reverse bias raises the barrier and decreases the diffusion.

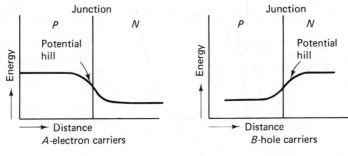

FIGURE 1-10. Energy Graphs for Common Diodes

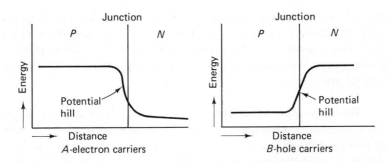

FIGURE 1-11. Energy Graphs for Tunnel Diodes

The tunnel diode *does not* behave in accordance with this diffusion process. Examine the energy graphs in Fig. 1-11 and compare them to the graphs in Fig. 1-10.

Notice that the tunnel diode has a much higher and steeper potential hill than the common diode. We would naturally conclude from this graph that a much greater energy would be required in order to lift the carriers over the hill. But such *is not* the case. In fact, when we apply a small forward bias, we obtain a strong forward current. As we increase the forward bias, another strange thing happens; we expect the current to increase, but it decreases instead. The carriers suddenly appear on the other side of the barrier as if they tunneled through the barrier. We know that they did not have sufficient energy to climb over the potential hill. In the tunneling process, they have developed a negative resistance in our diode. This is further described by the characteristic curve in Fig. 1-12.

A forward bias of 50 to 60 mV is sufficient to raise the forward current to the peak at point A. This is the point where tunneling begins and the negative resistance area is encountered. As we further increase the forward bias, the current gradually decreases to the valley at point B. This valley generally occurs when our forward bias

14 Solid State Devices

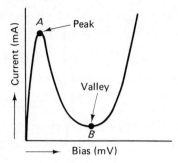

FIGURE 1-12. Characteristic Curve of Tunnel Diode

is about 350 mV. As we continue to increase forward bias beyond point B, the current increases in a normal manner, similar to other diodes. We normally operate our tunnel diode on the linear portion of the negative resistance curve between points A and B. This is the tunnel area.

Reverse Characteristics The tunnel diode has a straight positive resistance characteristic in the reverse direction. A very small reverse bias will drive this diode into the breakdown region. However, reverse characteristics are of passing interest, since we buy a tunnel diode for some application of its tunneling characteristics.

Back Diodes Some tunnel diodes are designed to operate in reverse. We call them back diodes. These diodes exhibit their tunneling characteristics in the reverse bias region.

Schematic Symbols Designers and manufacturers have not yet reached an agreement on the standard schematic symbol for tunnel diodes. Some of the most used symbols are shown in Fig. 1-13.

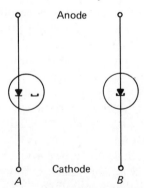

FIGURE 1-13. Schematic Symbols for Tunnel Diodes

The cathode and anode designations *are not* part of the symbol. Symbol A is the preferred symbol, but you will likely encounter symbol B from time to time.

Applications This rugged little diode is often smaller than the head of a pin, but it is very large in application. We frequently find it used as a high speed switch, an amplifier, and an oscillator. It can switch more than 100 times faster than a bipolar transistor, and its frequency capabilities commonly exceed 100 MHz. It can handle current in excess of 5 A.

ZENER DIODES

The Zener diode is named for C. Zener, who advanced the quantum mechanics theory of electrical breakdown in solids. We find that once breakdown occurs in a Zener diode it rapidly develops into a virtual avalanche of reverse current. This destroys the common diode, but the Zener diode is designed to operate on this effect.

Characteristics Like any other diode, as we increase the reverse bias on a Zener, the reverse current gradually increases. But suddenly we reach the breakdown potential, and the current avalanches along a line that is almost perpendicular. We are speaking of only a few microamperes, but this is reverse current. This avalanche characteristic is illustrated in Fig. 1–14. The point of breakdown, A on the graph, is the start of the vertical portion of the current curve. We call this vertical portion the *Zener region*, and the center of the Zener region is the proper operating current.

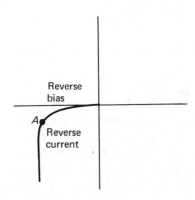

FIGURE 1–14. Zener Current-voltage Curve

16 Solid State Devices

Applications We find that one of the most popular uses of the Zener diodes is in voltage regulators. In this application, the diode parallels the load resistor in a regulated power supply. Minor changes in load current are routed through the Zener diode. This keeps a constant current through the load resistor and a steady dc voltage output. Figure 1–15 is the schematic symbol for the Zener diode. The cathode-anode designations *are not* part of the symbol.

FIGURE 1–15. Schematic Symbol for the Zener Diode

Symmetrical Zener The symmetrical Zener is a more recent expansion of the Zener principle. In effect, it is two Zener diodes connected back to back. The result is that we can have Zener action (avalanche breakdown) in either direction. It has no forward side, and bias in either direction drives the diode into the Zener region. The current-voltage curve and the schematic symbol are shown in Fig. 1–16.

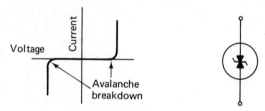

FIGURE 1–16. Symmetrical Zener

SILICON CONTROLLED RECTIFIERS

The silicon controlled rectifier (SCR) is a four-layer, solid state switch with three terminals. The term "rectifier" is misleading in this case, because we seldom use the SCR for anything except an electronic switch. But this still leaves us with many practical applications. Among other things, we can use the SCR as a light dimmer, an on-off switch, an amplifier, a time delay, overload protection, and a light flasher.

Symbol and Structure The schematic symbol and structure of the SCR is illustrated in Fig. 1–17. In addition to anode and cathode, we now have a third terminal, which we call a gate. The name of each terminal *is not* part of the symbol. The four-layer structure may be the PNPN as illustrated or a reverse order (NPNP).

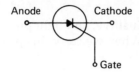

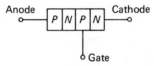

FIGURE 1–17.   SCR Symbol and Structure

Characteristics When we have *no input* at the gate, the switch is off. The anode current at this time is only slightly more than leakage current. We can turn the switch on by applying a pulse of current to the gate. This results in an increase in anode current and a strong regenerative feedback which drives the diode into saturation. The switch is on and the gate has no further control. The only limit to current is the external circuit. We shut off the switch by reducing the anode current. When this current drops to a critical level, which we call *holding current*, the diode switches to the off condition.

In operation the SCR functions as two transistors: one an NPN and the other a PNP. When arranged as shown in Fig. 1–18, these transistors function similar to an SCR.

The SCR is marketed under many names. We find them

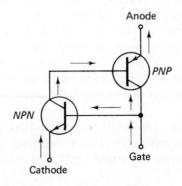

FIGURE 1–18.   Equivalent SCR Circuit

18 Solid State Devices

called silicon controlled switches, gate turn-off switches, light-activated silicon controlled rectifiers, Shockley diodes, and still other names.

PHOTODIODES

Many SCRs are light sensitive, and light provides the trigger current to operate the switch. When we carry this principle one step further, we have the photodiode.

Structure The photodiode has its PN junction exposed to a light source. The light controls the barrier resistance and, in turn, the quantity of current. The photodiode structure is illustrated in Fig. 1-19.

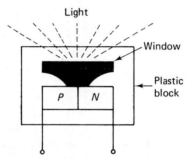

FIGURE 1-19. Photodiode Structure

Characteristics The photodiode has a window that focuses light on the PN junction. We have the junction reverse biased when the window is in darkness. The reverse current is on the order of 1 to 3 μA. When exposed to light, the same bias produces a much larger current. A small change in light intensity produces a very noticeable change in the reverse current. If we place forward bias on the junction, the light causes no appreciable change in the level of current. We may compare the photodiode to a potentiometer. As light intensity increases, reverse resistance decreases to cause an increase in current. For maximum sensitivity, we should aim the lens toward our light source. Sometimes we may need a tubular shield to protect the lens from stray light sources.

Applications We find the photodiode used for reading tapes and cards in computer data processing systems, automatic reading systems, automatic night light systems, and counting and detecting circuits.

INTRODUCTION TO TRANSISTORS

The first transistor was constructed at the Bell Laboratories in 1947. This point contact transistor was rapidly displaced by a three-layer transistor, which is in use today. For many years, a transistor was simply that: a transistor. The fairly recent development of the field effect transistor caused us to become more definitive in our nomenclature.

BIPOLAR TRANSISTORS

A bipolar transistor is basically two diodes placed back to back. Not two physically separate diodes, but two diodes within a single piece of crystal.

Types The bipolar transistor comes in two types: the NPN and the PNP. Both types are created by separating two like materials with a thin section of unlike material. The two junctions are formed as previously described under diodes. Both types of transistor are illustrated in Fig. 1-20. Notice that we now have two types of material with two junctions. This gives us a solid state device that not only is capable of amplifying a signal, but does a very fine job of amplification when it is properly connected.

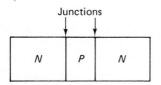 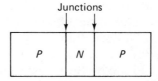

FIGURE 1-20. Types of Transistors

Transistor Elements The three sections of the bipolar transistor are the three elements as illustrated in Fig. 1-21. The *emitter* injects current carriers into the circuit, the *collector* collects these carriers, and the *base* controls the amount of current that is allowed to pass through the transistor. In a broad sense, we may compare these three elements to the cathode, anode, and control grid of a triode electron tube. However, such a comparison must be used with caution. The electron triode is a voltage sensitive device, whereas the bipolar transistor is a current sensitive device. Later we shall discuss the field effect transistor. It is a voltage sensitive device,

20 Solid State Devices

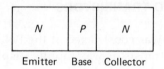

FIGURE 1-21. Elements of a Transistor

and it can do many things beyond the capability of the electron tube triode as well as the bipolar transistor.

Back to the bipolar transistor: the center section is always the base which puts it in a position of control. If the two outer sections are perfectly symmetrical, they may be arbitrarily designated as emitter and collector. In most cases, the two outer sections *are not* the same size. This means that the section designated as the emitter and the section designated as the collector were designed for that specific function, and they *are not* interchangeable.

Biasing the Junctions Each of the junctions in the bipolar transistor will permit current in one direction and block it in the other direction. Each junction has its own biasing circuit with the proper polarity to aid current through the junction. The proper biasing circuits for both NPN and PNP transistors are illustrated in Fig. 1-22.

Notice that in both cases the input junction (emitter-base) is forward-biased. That is, the bias polarity aids the movement of current carriers through the junction from emitter to base. The output junction (collector-base) is reverse-biased, which aids the movement of current carriers through the junction from base to collector.

Remember that current carriers may be either electrons or holes. In the NPN, we have electrons for carriers. We bias the input junction with a polarity that will cause electrons to be injected into the base from the emitter. Now we must bias the output with a polarity that will pull these electrons through the base into the collector.

In the PNP, our current carriers are holes, and the holes move opposite to the flow of electrons. We bias the input junction in a manner that will cause holes to be injected from the emitter into the base. Now we must bias the output junction in a manner that will pull the holes through the base into the collector.

When the bias polarities are as shown in Fig. 1-22, we find that current, both through the transistor and in the external circuit, is in the direction indicated by the polarity of the batteries. Notice that

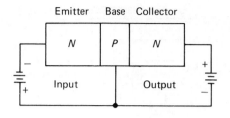

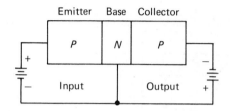

FIGURE 1-22. Proper Biasing Polarities

the polarities of both batteries are reversed when we go from one type of transistor to the other.

Current Direction and Division We find that the total circuit current passes through the emitter. Therefore, emitter current and total circuit current are one and the same. This emitter current is composed of collector current and base current. The collector current represents about 98 percent of the total, which leaves only about 2 percent for the base current. The direction and division of the currents are illustrated in Fig. 1-23 for both types of transistor.

In the NPN, out total current enters through the emitter and then divides between the base and the collector. About 98 percent of this total current exits through the collector. The remaining 2 percent comes out via the base lead.

In the PNP, the current enters from two sources. About 98 percent enters through the collector, and the remaining 2 percent enters through the base. These combine for total current and emerge through the emitter.

Applications The pressing need for a better amplifying device was the impetus behind the invention of the transistor. Therefore, we should expect to find the bipolar transistor in almost any type of circuit that requires amplification of either voltage, current, or power.

22 Solid State Devices

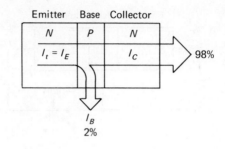

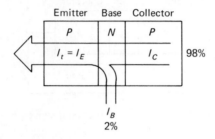

FIGURE 1-23. Transistor Currents

Schematic Symbols The schematic symbols for both NPN and PNP bipolar transistors are shown in Fig. 1-24. The NPN and the PNP work equally well in any application. The only difference is the polarity of the two bias voltages. All bias voltages must be reversed when we change from one type of transistor to the other. The element with the arrow is always the emitter, and the forward current is against the direction of this arrow. When the arrow points toward the base, it is PNP. When the arrow points away from the base, it is NPN.

FIELD EFFECT TRANSISTORS

The field effect transistor (FET) differs from the bipolar transistor as much as the tunnel diode differs from the common rectifier diode. We sometimes refer to this FET as an unipolar or unijunction transistor, because it has only one junction. The basic FET is properly called a junction field effect transistor (JFET). The other more recent type of field effect transistor is made from metal oxide. It is called a metal oxide semiconductor field effect transistor (MOSFET). We sometimes refer to the MOSFET as an insulated gate field effect transistor (IGFET).

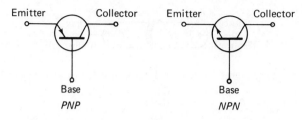

FIGURE 1-24. Schematic Symbols for Bipolar Transistors

Junction Field Effect Transistors (JFET) We can create a JFET by taking a single crystal of either P- or N-type material and diffusing a ring of different material around it, as shown in Fig. 1-25(a). The source, drain, and gate correspond to emitter, collector, and base of a bipolar transistor. The cutaway in Fig. 1-25(b) shows how the potential on the ring (gate) can be used to control the current between the source and the drain.

When we place a potential between source and drain, but have zero voltage on the gate, the current from source to drain sets up a reverse bias along the surface of the gate (junction). As we increase the drain-source voltage, the reverse bias increases until the drain-source current levels off at a maximum value. The drain-source voltage that causes this current limiting condition is called the pinch-off voltage V_p. This condition is similar to vacuum tube saturation. Further increases in voltage have very little effect on current.

When we desire to exercise control of the current between the source and the drain I_p, we place a negative potential on the gate. This potential need only be negative with respect to the source potential. As we increase the negative gate potential, we effectively increase the internal resistance between source and drain. The greater the negative gate potential, the higher the source-drain potential necessary to obtain a given level of drain current. Actually, the internal action is a result of an electric field that builds up around the gate (junction). This field repels the carriers and constricts their passage. The stronger the gate potential, the stronger the field, the greater the constriction, and the smaller the current that is allowed to pass.

We find JFETs in both N channel and P channel. The type of channel (N or P) refers to the crystal body between the source and the drain. The opposite type of material is, of course, used for the gate. In bipolar transistors the NPN and PNP used exactly reverse polarities on all bias voltages. In the FET, we must also reverse the bias when changing from an N channel FET to a P channel FET and

24 Solid State Devices

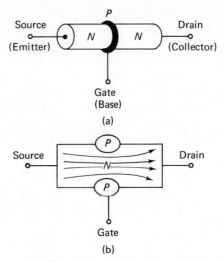

FIGURE 1-25. Construction of JFET

vice versa. With the N channel, we use either a positive potential at the drain or a negative potential at the source, and the drain must be positive with respect to the source. With the P channel, our source must be positive with respect to the drain. Figure 1-26 illustrates the schematic symbol and the proper bias for both P and N channel JFETs.

Notice the direction of current and the direction of the arrow on the gate. The symbol indicates that the JFET is a device with three leads, and the gate arrow points inward for N channel FETs and outward for P channel FETs. In both cases, the gate functions similarly to the control grid in a vacuum tube. The signal is applied to the gate, and the drain current is increased and decreased with the signal variations.

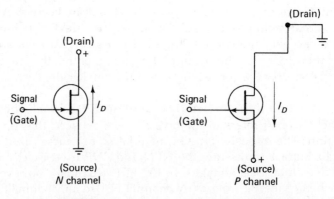

FIGURE 1-26. Symbology and Biasing of JFET

MOS Field Effect Transistors The metal oxide semiconductor field effect transistor (MOSFET) differs from the JFET in construction, especially the control mechanism, yet the two types of FET are interchangeable in many of our applications. We construct a MOSFET by diffusing two separate, low resistive regions of either P- or N-type material into a substrate of different material. We now cover the surface with an insulating oxide. See Fig. 1–27(a). The drain is constructed by etching a hole through the insulation and connecting a lead to one of the diffused regions. The source is constructed by the same process with the other diffused region. We now have the partial FET illustrated in Fig. 1–27(b). This is a substrate, two diffused regions, and leads for source and drain. The gate lead is connected to a strip of metal, which is placed across the gap between the source and drain regions. We must stress the fact that we have no electrical contact between the metal gate and any of the crystal material. We now have the completed MOSFET as illustrated in Fig. 1–27(c). Apparently we have two isolated regions of N-type material (this example) in a substrate of P-type material. The gate appears isolated from everything. This is substantially true when we have no signal on the gate, but, as we will see, this changes dramatically under operating conditions. The MOSFET in this illustration is an N channel MOSFET. If we reversed the types of material, it would then be a P channel MOSFET.

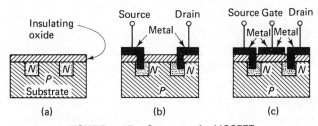

FIGURE 1–27. Structure of a MOSFET

Now let us examine a MOSFET with operational potentials applied to the elements. The bias voltages are the same as those described for JFETs; the drain of the N channel is more positive than the source, and the drain of the P channel is negative with respect to the source. We shall consider an N channel MOSFET. When we apply a positive signal to the gate, a capacitive action between the metal gate and the P-type substrate induces a conduction channel between the source and the drain regions. We call this channel enhancement because the positive voltage on the gate causes redistribution of the carriers in the substrate. For practical purposes, this induces a section of N-type material which connects the source region to the drain region. This action is illustrated in Fig. 1-28. Of course, if we were dealing with a P channel MOSFET, the channel would be induced by a negative gate potential, and the induced channel would be essentially P-type material. We call this type of operation, in either N or P channel MOSFETs, enhancement mode operation or channel enhancement.

Another method of constructing a MOSFET is to diffuse a low resistivity channel between the source and the drain regions. The gate then controls the distribution of carriers in the diffused channel. This type of MOSFET can be designed to operate in the channel depletion mode. The channel depletion mode of operation functions similar to reverse bias; the gate potential reduces the carriers in the conduction channel. In either case, the gate potential regulates the resistance of the conduction channel and controls the level of drain current.

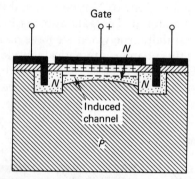

FIGURE 1-28. MOSFET with Induced Channel

Unijunction Transistor We have a transistor which is referred to as a unijunction transistor (UJT). Its construction is somewhat similar to a junction field effect transistor (JFET). We find that they differ principally in the size of the junction area. The emitter junction area of our UJT is much smaller than that of the JFET. The UJT is a three-terminal device with only one junction. It consists of two bases and an emitter. Figure 1-29 illustrates the structure, schematic symbol, and equivalent circuit of the UJT.

We sometimes refer to the unijunction transistor as a double-base diode because of its two bases. The example represented here is a P-type emitter with N-type bases. If we reverse the materials, we also reverse the emitter arrow. In the equivalent circuit, R_{B1} is represented as a variable resistor. This resistance value is a function of the bias voltages. The two principal currents are I_{B2} and I_E. The currents are controlled by the resistance of B1.

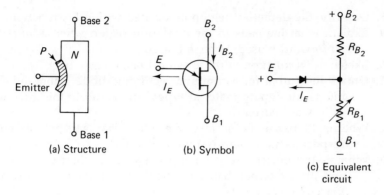

FIGURE 1-29. Unijunction Transistor

REVIEW EXERCISES

1. Name the three categories of materials.
2. In terms of electrons, what is the primary difference between conductors and insulators?
3. What is a semiconductor material?
4. Compare the three categories of materials in respect to resistivity.
5. What two materials are most often used to manufacture solid state devices? What type of structure do they have?
6. A stable atom with a full valence band has what number of electrons in its outer band?
7. Describe the covalent bonding of pure silicon and germanium.
8. What is the valence structure of trivalent and pentavalent materials?
9. Describe the physical structure of materials which have: negative current carriers; positive current carriers.
10. What is the difference between donor materials and acceptor materials?
11. Starting with a melt of pure silicon or germanium, what can be done to make donor material from this melt?
12. Starting with a melt of pure silicon or germanium, what can be done to cause the melt to produce acceptor materials?
13. What is the difference between P- and N-type materials?
14. Define electron current and compare it with conventional current.
15. Describe the movement of holes and electron current when a battery is connected across a crystal of P-type material.
16. Describe the movement of carriers and electron current when a battery is connected across a crystal of N-type material.
17. How is a solid state diode formed?

28 Solid State Devices

18. Describe the depletion region associated with a PN junction.
19. Explain what happens to the depletion region when a junction is: forward biased; reverse biased.
20. Sketch a PN junction with: forward bias; reverse bias.
21. Draw the symbol for a common rectifying diode. Label the elements, designate material types, and indicate the direction of forward current.
22. What is the barrier potential in a diode? What is its polarity with respect to the P- and N-type materials?
23. What is the polarity relationship between the barrier potential and the external battery when the junction is: forward biased? reverse biased?
24. What happens to the magnitude of the barrier potential when the external bias is increased in the: forward direction? reverse direction?
25. Describe the internal action of a diode when it is subjected to a level of reverse bias beyond the limits of the barrier potential.
26. What is the effect of avalanche breakdown on a common diode?
27. What is the relationship between junction temperature and current carriers?
28. Describe the condition known as thermal runaway. What is the probable result?
29. What is the primary difference between a signal diode and a power supply diode? What are the reasons for this difference?
30. When the temperature *is not* specified in a diode rating, what temperature must be assumed?
31. What is indicated by the WVDC rating?
32. Interpret this signal diode rating: $B_v = 45$ V.
33. What is indicated by a PIV rating?
34. What information is given by an I_f rating?
35. Draw a current-voltage curve for a tunnel diode. Label the peak, valley, negative resistance area, and the operating point.
36. Tunneling takes place during what portion of the current-voltage curve?
37. Draw two symbols for a tunnel diode, and indicate anode, cathode, and direction of tunnel current.
38. Draw the current-voltage curve for a Zener diode and indicate breakdown voltage, Zener region, and operating point.
39. Draw the symbol for a Zener diode and indicate anode, cathode, and direction of avalanche current.
40. Draw the characteristic curve and symbol for a symmetrical Zener.
41. Draw a sketch to represent an NPN transistor showing both junctions. Draw batteries to bias both junctions for proper operation. Label the three elements.

42. Draw the schematic symbol for an NPN and a PNP transistor. Label the elements. Indicate the direction and percentage of total current for each element.
43. Describe the bias arrangement and control action of a JFET.
44. What is the difference between a P channel and an N channel JFET?
45. What are the meanings of the letters MOSFET?
46. What is meant by "induced channel" in a MOSFET?
47. Draw a schematic symbol for a P channel and an N channel FET. Label the elements and indicate the direction of I_D
48. What is meant by a "diffused channel" in a MOSFET?

Chapter Two

Energy Conversion Circuits

In this chapter we shall analyze circuits that are repeated in many types of electronic power supplies. All these circuits use diodes, so we shall concentrate heavily on diode applications. However, the primary objective is to attain a clear understanding of the over-all purpose and function of each circuit. The diode now becomes one of the circuit components and assumes a secondary role. A few of the circuits also utilize transistors as circuit components.

Some form of energy conversion circuits is required in nearly all electronic equipment. The primary reason for this requirement is the method of generating and delivering voltage to the user. This voltage is produced and delivered in the form most convenient to the supplier. The most convenient form, at least in the United States, is alternating voltage at a frequency of 60 Hz. If any part of our electronic equipment requires anything except a 60-Hz alternating voltage, we must convert this voltage to fit our own needs.

POWER TRANSFORMERS

A power supply for equipment using solid state devices must deliver a relatively high current with a low voltage. These requirements must be considered in the power supply design. The transformer is the first circuit component that we encounter.

Our transformer must be designed specifically for use with solid state circuits. This means that it must have relatively few turns on the secondary to enable it to deliver a low voltage. At the same time, the cross-sectional area of the wire must be larger in order to carry the high current that is needed. The large wire reduces the resistance and keeps the transformer power losses to a minimum.

Power frequencies range from 60 to 1600 Hz, but a particular transformer is designed to handle only one of these frequencies. The most common is 60 Hz, and this is generally a single-phase voltage. However, we do encounter triple-phase voltage quite often. The amplitude of the input voltage is usually 120 V. Since we are speaking of alternating voltage, that figure is understood to be the effective (root mean square) value.

Transformer Efficiency Theoretically, a perfect transformer has a one-to-one ratio between the input power to its primary and the output power that it delivers from its secondary. Of course, we have no perfect transformers, but we do have transformers that deliver an output power that is more than 99 percent of the input power. We calculate efficiency of a transformer by dividing the power out by the power in. Since efficiency is normally given in percentage, we multiply the result of this calculation by 100. When we express this as an equation, we have

$$\% \text{ efficiency} = \frac{P_{out}}{P_{in}} \times 100$$

The transformer in Fig. 2-1 has a primary power of 48 watts and a secondary power of 47 watts. What is the efficiency of this transformer?

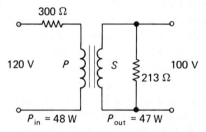

FIGURE 2-1. Power In versus Power Out

$$\% \text{ efficiency} = \frac{P_{out}}{P_{in}} \times 100$$

$$= \frac{47 \text{ W}}{48 \text{ W}} \times 100$$

$$= 0.979 \times 100$$

$$= 97.9 \text{ percent}$$

This transformer is 97.9 percent efficient. What became of the one watt of power that we lost between the primary and the secondary?

Energy Losses There are three kinds of energy losses in a transformer: copper losses, hysteresis losses, and eddy current losses. Copper wire has a very low resistivity, but a given length of wire does have a small resistance value. This resistance causes heating as electrons flow through the primary and secondary windings. For instance: a winding may have a resistance of 1Ω and a current of 1 A. This winding would have a copper loss of

$$\begin{aligned} \text{Copper loss} &= I^2 R \\ &= (1 \text{ A})^2 1\Omega \\ &= 1 \text{ W} \end{aligned}$$

Power transformers are constructed with iron alloy cores, and the iron cores account for hysteresis losses and eddy current losses. When we magnetize a piece of iron, a small quantity of magnetic flux will remain in the iron after we remove the magnetizing force. We call this residual magnetism, and this residual magnetism is one of our problems. The core in a power transformer is energized in opposite directions with each alternation of the input voltage. Each time the voltage waveshape crosses the zero point, the residual magnetism must be cancelled before the magnetic field can be reversed. If we take the quantity of energy used to cancel the residual magnetism and multiply it by two, we have the hysteresis loss for each cycle of the input voltage. With our modern alloys, transformer cores can be constructed with very little residual magnetism, but it is always present and always causes some losses due to this hysteresis effect.

Anytime we place a conducting material in a position where it will be exposed to a changing magnetic field, alternating current is induced into the conductor. The iron core in our transformer is a conducting material, and it is continuously exposed to a changing magnetic field. Each time the current in the winding builds up and dies out, the magnetic field expands and decays through the iron core. This changing field induces alternating current into the transformer core. We call these currents eddy currents. The eddy currents heat the core, and this represents a loss of energy.

In power transformers, we have only one way to fight the eddy currents. We must make the path for these currents as short as we possibly can. The paths can be shortened by laminating the iron used in the core. The lamination process involves slicing the iron core into thin slices and coating each slice with an insulating varnish. When the slices are assembled, we still have an iron core on which to wind the coils, but eddy currents have only very short paths within the iron. The heating effect is greatly reduced, and the losses are very small. Figure 2–2 illustrates one pattern of core lamination and the method of winding the wire on the laminated core. We call this pattern O ring lamination.

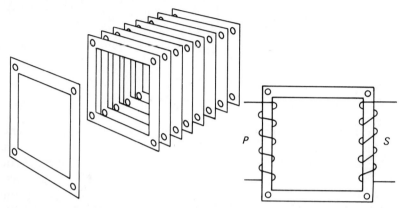

FIGURE 2–2. Lamination and Winding with O Ring Pattern

Transformer Ratios We generally have 120 volts delivered on the power line; our transformer must step it down to a usable level. In this case, we want 20 V dc. This would seem to indicate that we need a 6 to 1 stepdown ratio, but this *is not* exactly the case. The 120 V is the alternating effective value, and the 20 V is a direct voltage value. However, we must obviously use a stepdown transformer and reduce the input voltage by some amount. The exact amount of stepdown will be determined by the type of rectifiers and regulators that we use between the transformer and the solid state circuits to which we are supplying voltage.

It is an easy matter to design a transformer that will step the voltage down while stepping the current up. It is a matter of placing fewer turns of wire on the secondary than we have on the primary. The exact relationship between turns and voltage is expressed by this equation:

$$\frac{E_p}{E_s} = \frac{N_p}{N_s}$$

where E_p = voltage across the primary, E_s = voltage across the secondary, N_p = turns on the primary, and N_s = turns on the secondary.

The transformer in Fig. 2–3 is a 5:1 stepdown transformer.

34 Energy Conversion Circuits

What is the voltage across the secondary? The primary of this transformer has 100 turns of wire. How many turns are in the secondary winding?

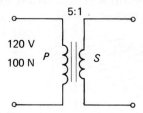

FIGURE 2-3. Stepdown Transformer

The primary voltage is 120 V, so the secondary voltage must be

$$\frac{120 \text{ V}}{5} = 24 \text{ V}$$

The turns ratio is 5:1, so with 100 turns on the primary, the secondary turns must be

$$\frac{100}{5} = 20 \text{ turns}$$

As you can see, we have a direct relationship between the primary to secondary turns ratio and the primary to secondary voltage ratio. The actual number of turns on the two windings is not important; it is the ratio between primary turns and secondary turns. What is the turns ratio of the transformer in Fig. 2-4? How many turns of wire does the primary winding contain?

The voltage ratio is

$$\frac{E_p}{E_s} = \frac{100 \text{ V}}{10 \text{ V}} = \frac{10}{1}$$

or

10:1

Since

$$\frac{E_p}{E_s} = \frac{N_p}{N_s}$$

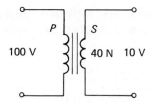

FIGURE 2-4. Calculate Ratio and Turns

The turns ratio must also be 10:1. Since the secondary contains 40 turns, the primary must contain 10 × 40, or 400 turns.

The same turns ratio which steps down the voltage between the primary and secondary also steps up our current. Let us go back to the previous illustrations; the transformer in Fig. 2-3 has five times as much current in the secondary as it has in the primary. The transformer in Fig. 2-4 steps the current up by a multiple of 10 between the primary and the secondary. By these facts, we can see that there is an inverse relationship between the turns ratio and the current ratio. We express these ratios with this equation:

$$\frac{I_s}{I_p} = \frac{N_p}{N_s}$$

These ratios are all very convenient for solid state power supplies. The voltage from the power line is always higher in amplitude than we need, but it produces a small current in the primary. The transformer is designed to reduce the voltage to a usable value while raising the current to the desired level. Both of these objectives are easily obtained by using a stepdown transformer with a specific turns ratio.

Since the impedance of a transformer primary is affected by the impedance of the secondary, there must also be a direct relationship between the turns ratio and the impedance ratio. The following equation expresses that exact relationship:

$$\frac{Z_p}{Z_s} = \left(\frac{N_p}{N_s}\right)^2$$

Transformers are available with a wide variety of turns ratios. This selection of ratios makes it easy to match the impedance of the previous circuit to the primary winding while matching the impedance of the secondary to the circuit the transformer is feeding. This is a very important consideration, because maximum power transfer takes place when the impedances are so matched.

The transformer in Fig. 2–5 has a matched impedance at both the primary and secondary. What is the turns ratio? What is the power in the primary? In the secondary?

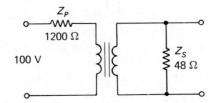

FIGURE 2–5. Turns Ratio and Power Transfer

The impedance ratio is equal to the square of the turns ratio:

$$\frac{Z_p}{Z_s} = \left(\frac{N_p}{N_s}\right)^2$$

$$\frac{1200}{48} = \left(\frac{N_p}{N_s}\right)^2$$

$$\frac{N_p}{N_s} = \sqrt{25}$$

$$\frac{N_p}{N_s} = 5 \text{ or } 5:1$$

The power of the primary is

$$P = \frac{E^2}{R}$$

$$= \frac{100 \times 100}{1200}$$

$$= \frac{10{,}000}{1200}$$

$$= 8.33 \text{ W}$$

The power in the secondary is

$$P = \frac{E^2}{R}$$

$$= \frac{20 \times 20}{48}$$

$$= \frac{400}{48}$$

$$= 8.33 \text{ W}$$

Primary to Secondary Phase Relation When dealing with direct voltages, we compare polarities: positive and negative. Since alternating voltage changes polarity in a periodic fashion, it is better to compare phases. Voltage may be delivered to our transformer as single, double, or triple phase. A matching transformer may be selected to handle either type of phase input. A transformer with a two-wire primary, such as those in our previous illustrations, is intended for single-phase inputs. The two leads from the secondary winding indicate that the transformer also has a single-phase output.

A primary with three leads is generally used to handle a double-phase input. The two voltages of a double-phase voltage are usually equal in amplitude and 180 degrees out of phase. If they are exactly balanced, the ground wire has zero current and may be eliminated. The transformer to handle double-phase voltage is illustrated in Fig. 2–6.

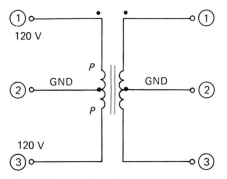

FIGURE 2–6. Double-phase Transformer

In Fig. 2–6 both primary and secondary windings are center tapped. This gives the effect of two separate windings on both primary and secondary. The ground wire is the return path in both cases, and the voltage at point 1 is 180 degrees out of phase with the voltage at point 3. This phase relation is true for both input and output. The dots on the transformer are phase indicators. These dots are not always used, but when they do appear, all points with dots have a common phase.

A transformer with only two leads on the secondary has a single-phase output regardless of the input. On the other hand, if our secondary winding has a center tap, we have a double-phase output regardless of the input.

Three-phase generators are common sources of alternating voltage, and three-phase transformers are needed to handle the resulting triple-phase voltage. There are two popular styles of triple-phase transformers: delta (Δ) and wye (Y). It is common practice to use one

38 Energy Conversion Circuits

style for the transformer primary and the other style for the secondary. Figure 2-7 illustrates a transformer with a Δ primary and a Y secondary.

Regardless of style, the illustrated transformer has three phases in and three phases out. Each voltage is 120 degrees out of phase with both voltages on adjacent windings. No ground wire is shown, and if the voltages are perfectly balanced the ground wire is not necessary. With the Y secondary, outputs may be taken between any two adjacent leads. In this case, an output becomes 1.73 times the voltage on either coil.

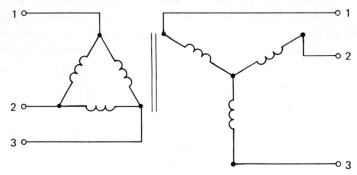

FIGURE 2-7. Triple-phase Transformer

POWER SUPPLY RECTIFIERS

An ideal rectifier would have zero resistance in the forward direction and infinite resistance in the reverse direction. Since we do not have this ideal rectifier, we must operate with something less. When selecting the best available rectifier for a particular power supply, we must consider the operating conditions that it will be exposed to. These conditions include voltage and current requirements, peak inverse voltage, possible transients, and cooling requirements.

Resistance Ratio We normally do not require a special, high-priced diode for a power supply rectifier. If we keep voltages and currents within the recommended limits, and avoid abuse, ordinary diodes will serve this purpose very well. Commercially available diodes have a very small forward resistance. The reverse resistance of these diodes is not infinite, but it is so large that we can ignore it. This is possible because the reverse diode resistance is a great deal larger than the load resistance.

Peak Inverse Voltage Sometimes the PIV rating of a single diode is less than the value we need for our power supply. This is a small problem. We can connect two or more of these diodes in series and increase the PIV rating accordingly. But we must keep in mind that two diodes in series will have twice the normal forward resistance at the same time that they will withstand twice as much reverse voltage. This increased forward resistance will cause some loss in voltage amplitude and an increase in the operating temperature. If the temperature is likely to approach the rated value, it is best to use diodes with heat sinks or cooling fins to help dissipate the extra heat.

Increasing Current Capacity We may also place diodes in parallel when a single diode has less than our required current capacity. Two diodes in parallel can handle twice as much current as a single diode. We must be careful in this case to assure that one of the parallel diodes is not required to handle most of the current while the other does very little work. To ensure an equal current through each parallel diode, we choose diodes with reasonably similar electrical characteristics.

Transient Considerations Our most serious transient problem occurs during switching actions, because switching causes sudden load changes. Opening or closing a switch can generate surges of voltage and current that are much greater than the normal amplitudes. If our power supply is subject to reasonably high surges, it may be advisable to choose either a germanium or a selenium diode for the rectifying element instead of a silicon diode. A good silicon rectifier has extremely low reverse current until it reaches the avalanche breakdown stage. Germanium and selenium rectifiers have more leakage in the reverse direction, which gives them a better chance to absorb the transients without damage.

RECTIFIER CIRCUITS

The purpose of a rectifier circuit is to change the alternating voltage from our transformer into a pulsating direct voltage. There are many types of rectifier circuits: half wave, full wave, bridge, etc. In solid state power supplies, we frequently use voltage multipliers in conjunction with the rectification process.

Half-wave Rectifier The half-wave rectifier has the simplest circuit of all rectifiers. However, it is quite adequate for many power requirements. This rectifier consists of a solid state diode in series with the transformer and the load resistance. The diode passes current during one alternation of each input cycle and blocks it on the other. This action produces a high peak of voltage on one half cycle and zero voltage on the other half cycle. Figure 2-8 illustrates the actions of a half-wave rectifier.

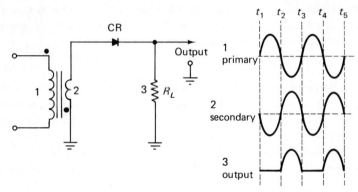

FIGURE 2-8. Half-wave Rectifier

Primary, secondary, and output waveshapes appear at points 1, 2, and 3 respectively. The rectifying diode conducts only on the positive alternation of the secondary voltage. During the negative alternation of the secondary voltage, the diode is reverse-biased and blocks the current. Electrons flow from ground, through the load resistor R_L, and through the diode during the half cycle when the diode is conducting. There is a negligible movement of electrons when the diode is cut off. This on-off action of the diode develops the output waveshape across the load resistor. These waveshapes are intended as voltage waveshapes, but they could just as well represent current. Our input alternating voltage has been changed to a pulsating direct voltage.

In a pulsating voltage, the frequency of the pulses may be an important factor. This frequency is especially important when the voltage needs to be smoothed out to a steady average value. With our half-wave rectifier, we have one pulse for each complete cycle of the input; so the pulse (or ripple) frequency is the same as the line frequency that is applied to the transformer primary.

The peak amplitude of the output voltage is only slightly less than the peak value of the secondary voltage. There is a very slight loss in amplitude because of the small forward resistance of the diode, but it is so small that we can ignore it. The average value of voltage amplitude during one alternation of a sine wave is 0.637 times the peak value. Since the half-wave rectifier uses only every second alternation, its average output amplitude is 0.637/2, or 0.318 times the peak amplitude.

Many solid state circuits require a relatively high voltage with very little current. The half-wave rectifier is well suited for such circuits. A simple capacitor filter can be used to smooth out the ripple, and with low current, a steady voltage is easy to obtain. We shall have more to say about filters later in this chapter.

The circuit illustrated in Fig. 2–8 is a positive rectifier. The electrons flow from ground through the load resistor, and as they do, a positive voltage is developed across the load resistor. In many cases, we require a negative voltage. This is a small problem. A negative output can be obtained by reversing the leads on the diode rectifier. This will reverse the direction of current through the load resistor, and the negative alternation will be preserved instead of the positive alternation.

When selecting a diode for use as a half-wave rectifier, we must remember that it will be subjected to a high peak inverse voltage. During the alternation when the diode is cut off, the voltage across the diode is equal to the peak-to-peak value of the voltage from the transformer. This amplitude is about 2.83 times the rms value of the transformer secondary voltage.

Current surges represent another hazard to the diode, and surges are present in all power supplies. One way to protect the diode is to connect a small resistor in series, as indicated in Fig. 2–9. Several factors affect the ohmic value of this surge resistor R_s. The type of filter used, the resistance of the transformer secondary, and the type of switching are some of the important factors. After all factors have been considered, the surge resistor value must be such that it will dampen the surges to tolerable values.

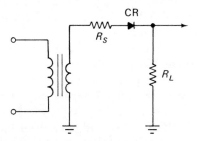

FIGURE 2–9. Placement of Surge Resistor

Full-wave Rectifier We find that the full-wave rectifier is the most popular type of rectifier in solid state power supplies. The most common arrangement that we encounter consists of a transformer with a center tapped secondary winding, two rectifying diodes, and a load resistor. These components are arranged as illustrated in Fig. 2–10. This is a positive rectifier with output voltage as indicated. If we reverse the leads to both diodes, the circuit becomes a negative rectifier, and the output waveshape is the same except that it is below the zero reference line.

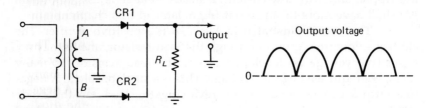

FIGURE 2–10. Full-wave Rectifier

During one alternation of our input voltage, point A is positive and electrons flow from ground, through the load resistor, through CR1 to ground at the center tap of the transformer. This current develops a positive pulse of voltage across R_L. During the next alternation of the input voltage, point B is positive and electrons flow from ground, through R_L, through CR2 to the same center tap ground. Another positive pulse of voltage is developed across R_L during this second alternation. Each pulse of the output voltage has an amplitude equal to half the peak value of the secondary voltage. So, for a given transformer, the output peaks from a half-wave rectifier are twice the amplitude of the peaks from a full-wave rectifier.

Each of our diodes conducts for one alternation and cuts off for one alternation. Only one diode is conducting at any given time. During its cutoff time, each diode is subjected to a reverse voltage equivalent to half the peak voltage of the transformer secondary. The ripple frequency is twice the frequency of the line voltage input, and this higher frequency keeps a relatively steady current through the load resistor. The output voltage from our full-wave rectifier is easy to filter and regulate. The relatively low voltage and high current characteristics of the full-wave rectifier, along with ease of filtering, make it suitable for nearly all solid state power supplies.

Bridge Rectifier Our bridge rectifier incorporates the good features of both half- and full-wave rectifiers. It has the full wave and high current characteristics of the full-wave rectifier as well as providing the high voltage of a half-wave rectifier. At least four solid state diodes are required, and we find that the circuit is arranged similar to the diagram in Fig. 2–11.

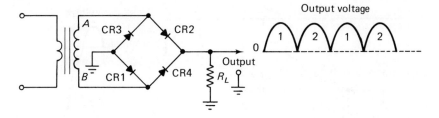

FIGURE 2-11. Bridge Rectifier

When point A is positive, CR1 and CR2 are forward-biased, and at the same time, CR3 and CR4 are reverse-biased. Electrons flow from point B through CR1 to ground, through R_L, through CR2 to point A on the transformer. During this alternation (point A positive), output 1 is developed across R_L. During the next alternation, point B is positive. Electrons now flow from point A through CR3 to ground, from ground through R_L, through CR4 to point B on the transformer. During this second alternation (point B positive), output 2 is developed across R_L.

The entire secondary of our transformer is used with each alternation. So, the amplitude of the output voltage is the same as the peak amplitude of the voltage on the secondary winding. Since each alternation produces an output pulse, the ripple frequency of the output is twice the frequency of the input voltage. As indicated in the illustration, the output from this rectifier is positive. If a negative voltage is desired, we simply reverse the connections on all four diode rectifiers. Since this arrangement places two diodes in series at any given time, we have no problem with peak inverse voltage. Each diode is subjected to a reverse bias of only the peak amplitude of our secondary voltage.

Three-phase Rectifier When we have a three-phase input voltage, we must have a three-phase power transformer. Since each secondary winding produces a sine wave of voltage, we need to rectify each of the sine waves. We can use half-wave, full-wave, or bridge rectification in either case. Regardless of the type of rectification employed, we normally collect the output across a single load resistor. If we choose half-wave rectification, our rectifier and waveshapes will be as illustrated in Fig. 2-12.

Waveshapes A, B, and C are developed across windings A, B, and C, respectively. Notice that these three waveshapes are separated by 120 degrees. Each of our rectifier diodes will conduct for a portion of the positive alternation of the voltage on the winding to

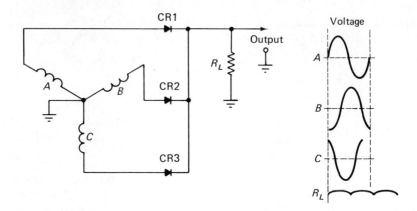

FIGURE 2-12. Three-phase Rectifier

which it is connected. This gives us an overlap in rectifier conduction time to such an extent that a high level of current is maintained through R_L at all times. The path for current is from the center of the wye to ground, from ground through R_L and through the two conducting diodes to the coils. The peaks of the output occur with each positive peak on the three coils. Therefore, our output ripple frequency is equal to three times the input frequency. This ripple is small in amplitude because of the overlap in rectifier conduction time. This is a positive rectifier, but we can change it to a negative rectifier by reversing the connections on all three diodes.

The peak inverse voltage on each diode in this arrangement is less than the peak-to-peak value on each coil but considerably more than the peak value. The PIV on a cutoff diode is approximately 2.45 times the rms value of voltage on a given coil. The amplitude of the output peaks is the same as the peak value on each coil.

POWER SUPPLY FILTERS

The pulsating direct voltage from our rectifiers is *not* suitable for use in most solid state circuits. Filtering is necessary to reduce the ripple amplitude and change the pulsations into a reasonably steady value of direct voltage. Of course, filter circuits will not do a perfect job, but they will get us a lot closer than we are when our voltage is taken directly from the rectifier.

Capacitor Input Filter The simplest filter that we can devise is a capacitor in parallel with the load resistor. Strangely enough, this is a very suitable filter for some applications. We are likely to find this type of filter when a half-wave rectifier is supplying voltage to a circuit that requires very little current. This resistor-capacitor (RC) filter is illustrated in Fig. 2–13.

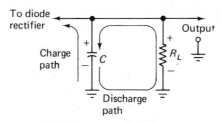

FIGURE 2–13. RC Filter

The capacitor charges when the diode rectifier conducts. Electrons flow from ground to the bottom plate of the capacitor and pile up on this plate. Electrons leave the top plate of the capacitor and flow through the rectifier. When the rectifying diode cuts off, the capacitor is charged with the polarity indicated. The capacitor attempts to discharge during the alternation that the rectifier is cut off. The only discharge path is through the load resistor, and this path has a very long time constant. The result is a very slight loss of charge before the rectifier conducts again.

The effectiveness of this filter is determined by the type of load placed across R_L. If the load draws a large quantity of current, it will discharge the capacitor, and very little filtering will be realized. If the load draws very little current, the capacitor will retain nearly a full charge, and the output will be a steady direct voltage.

When our load uses any appreciable current, we can use a coil to slow the capacitor discharge. The filter then becomes an LC filter, as illustrated in Fig. 2–14.

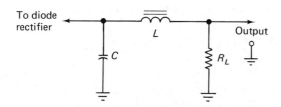

FIGURE 2–14. L-type LC Filter

This is an L-type, capacitor input, LC filter. The capacitor discharge and the decaying field on the coil combine to keep a steady value of current through R_L. The coil is effectively a short to dc and an open to ac. The capacitor is a virtual short for ac and an open for dc. The combined action of these two components provides effective filtering even when the circuits in parallel with R_L are drawing reasonable quantities of current. Coils designed for use in power supply filters use iron cores. The iron cores are possible because of the low frequencies of power supply voltages. The coils are very effective in blocking ac and all ac components of the ripple frequency. We commonly refer to these inductors as choke coils because of this ac blocking action.

When the load current requirements grow large enough to render the previous filters inadequate, we resort to the pi (π) type, capacitor input, LC filter. This filter is illustrated in Fig. 2–15. We are still using a half-wave rectifier; so current is still very limited, but this filter enables more current and still gives good filtering action. The charge and discharge action of the reactive components smooth out the slow changes while sudden changes are blocked by the coil and shorted by the capacitors.

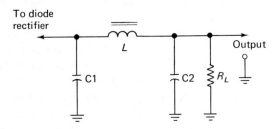

FIGURE 2–15. Pi-type LC Filter

The pi-type LC filter just described is also effective with full-wave rectifiers. When we use it with a full-wave rectifier, the capacitors charge twice as often and have only half the discharge time as with a half-wave rectifier. This enables us to furnish a much higher current while maintaining a reasonable degree of filtering.

Choke Input Filters When our output current requirements reach a certain critical value, the capacitor input filters become ineffective. When this happens, we change to a choke input filter. An L-type, choke input filter is effective for relatively large current drains. Such a filter is illustrated in Fig. 2–16.

This filter is used with a full-wave rectifier. As one of the diode rectifiers starts to conduct less, long before cutoff, the coil and capacitor react to the change and keep a steady current through R_L. The peak conduction of both rectifier diodes replenishes the charge on both reactive components. With very large current drains,

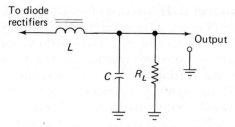

FIGURE 2-16. L-type Choke Input Filter

even this filter may fail to do an adequate job. When this happens, we can connect two or more L sections one after the other until we obtain the desired degree of filtering.

VOLTAGE MULTIPLIERS

At times we have need for extremely high voltages with solid state circuits. When the need does arise, voltage multipliers can be used to raise the level of our voltage to almost any desired value. When we use voltage multipliers in conjunction with our rectifiers, no further filtering is necessary. Devices that require very high voltages seldom use any appreciable current. This low current drain enables the voltage multipliers to function and reduces the need for elaborate filters.

Simplified Voltage Doubler The simplest form of a voltage multiplier is illustrated in Fig. 2-17. This is a modification of a fullwave rectifier circuit. It produces an output equal to twice the peak voltage on the transformer secondary.

On one alternation, point A is positive, and point B is nega-

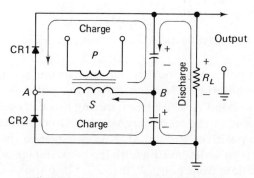

FIGURE 2-17. Simplified Voltage Doubler

tive. During this half cycle, electrons flow from point B, through C1, through CR1 to point A. This current charges C1, with polarity as indicated, to a level equal to the peak value of the voltage on the transformer secondary. On the next alternation, point B is positive, and point A is negative. During this half cycle, electrons flow from point A, through CR2, through C2 to point B. The current during this alternation charges C2, with polarity as indicated, to a level equal to the peak value of the voltage on the secondary of the transformer.

We now have each capacitor charged to the peak of the secondary voltage. The discharge path for each capacitor is through the load resistor and through the other capacitor. This discharge path has a very long time constant with respect to the input frequency. The charge of both capacitors is impressed across the load resistor. The output is the same as the voltage across R_L, and this is equal to twice the peak of the secondary voltage. The output is positive with respect to ground.

Cascade Voltage Doubler We can rearrange our capacitors and rectifier diodes and change the simplified voltage doubler to a cascade voltage doubler. The advantage of a cascade doubler is that it can be changed to a voltage tripler or quadrupler by adding sections. Figure 2–18 is a schematic diagram of a cascade voltage doubler.

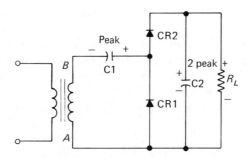

FIGURE 2–18. Cascade Voltage Doubler

When point A on our transformer secondary goes positive, point B goes negative. Electrons flow from point B, through C1, through CR1 to point A. The current during this half cycle charges C1 to the peak value of the secondary voltage. On our next alternation, point B goes positive, and point A goes negative. Now the peak charge on C1 is in series, aiding with the peak voltage on the secondary. Electrons flow from point A, through C2, through CR2, through C1 to point B. The current during this alternation charges C2 to twice the peak value of the secondary voltage. The attempt of C2 to discharge impresses this 2 × peak voltage across R_L with polarity as indicated.

Voltage Tripler We can convert the cascade voltage doubler to a voltage tripler by adding another section. This section will consist of a diode rectifier, a capacitor, and a resistor. We add this section to our doubler, as indicated in Fig. 2-19.

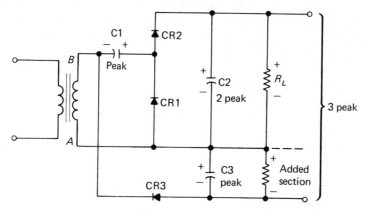

FIGURE 2-19. Voltage Tripler

In our added section, CR3 conducts on the same alternation as CR1. Current through CR3 charges C3 to the peak value of the secondary voltage. C3 discharges through the added load resistor, while C2 discharges through the original R_L. These two voltages in series are equal to three times the peak voltage on the transformer secondary.

VOLTAGE DIVIDERS

Solid state circuits require a variety of voltages. This includes voltages of different amplitudes as well as voltages of both positive and negative polarities. We handle the polarities with positive or negative rectifiers. Voltage dividers enable us to distribute the output voltage from a single rectifier circuit to satisfy several voltage levels. The total rectifier output must be equal to, or larger than, the largest load voltage. This amplitude requirement is easily obtained because most solid state circuits require relatively small values of voltage. Of course, there is a limit to the quantity of current that we can obtain from a given rectifier. This total current capacity must equal or exceed the total current to be drawn by all the circuits powered by the rectifier output.

Single Loads Any given circuit performs at top efficiency only when it has a specified voltage from a source that will supply the specified current. Obtaining proper voltage with the proper current is a very simple matter when we have only one circuit to supply. Two devices have been designed specifically for adjusting levels of voltage and current. The potentiometer is a variable resistor that is designed for voltage control; the rheostat is a variable resistor that is designed for current control. The simplest form of a voltage divider is a potentiometer connected as illustrated in Fig. 2–20.

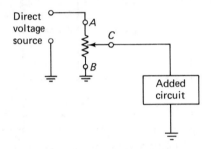

FIGURE 2–20. Potentiometer as a Voltage Divider

We frequently find potentiometers that can be adjusted from zero resistance to something in excess of 5 MΩ. Since potentiometers are expected to carry only a small quantity of current, they may be constructed from granulated carbon and have a very low wattage rating. The total resistance, point A to point B, is normally connected across the voltage source as shown in the illustration. This arrangement places the total applied voltage E_a across the potentiometer. The circuit to be powered is then connected to the movable arm of the potentiometer. The level of voltage that is delivered to the added circuit will be E_a - IR drop between point A and point C. We can vary this available voltage value from zero to the total value of E_a by moving the position of the movable contact. As long as E_a is equal to, or greater than, the voltage required by the added circuit, there is no problem. We simply move the control until the voltage between point C and ground is the desired value. Volume controls in various types of electronic equipment, and many other similar controls, are potentiometers used in the fashion just described.

The rheostat is a current control, so we generally find that it has a relatively small resistance value. Rheostats are normally composed of resistive wire that can withstand relatively large values of current. The rheostat is placed in series with the voltage source and the added circuit, as illustrated in Fig. 2–21. Here we are assuming that the voltage from the source is already very near the proper value. The variable arm may be connected to point B as shown here, but we may choose to leave point B open and connect the center arm to the circuit that is being controlled. These two connections provide the same results electrically. If we move the arm toward point A, we increase the current to the circuit. If we move the arm toward point B, we decrease the circuit current. Of course, these adjustments have a small effect on the voltage across the circuit, but this effect is small and of little consequence. This is not a voltage divider; it is a current control.

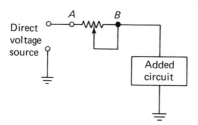

FIGURE 2–21. Rheostat as a Current Control

Multiple Loads The more circuits that we power from a single voltage divider, the more complex the divider. A voltage divider with no load is a simple series circuit with the applied voltage distributed across the divider resistor. Certain subquantities of E_a can be measured from the various taps to ground. This unloaded voltage divider is convenient from the standpoint of circuit simplicity, but it is a do-nothing circuit. The instant that we place a load on a voltage divider, we have a completely different picture. The divider circuit then becomes a series-parallel circuit, and each load affects the voltage distribution to all other loads. Let us reexamine the potentiometer voltage in Fig. 2–20. The added circuit is in parallel with part of the resistor: point B to point C. The remainder of the resistor is in series: point A to point C. This series portion of the resistor is carrying the total current. Each additional circuit that we add to the voltage divider further complicates this series-parallel arrangement, as illustrated in Fig. 2–22.

This diagram shows a voltage divider that is supplying voltage to three separate circuits. The series-parallel arrangement of circuits and voltage divider resistors causes an important interaction among the loads. The only way to have the proper voltage and the proper current through any load is to have all circuits functioning

52 Energy Conversion Circuits

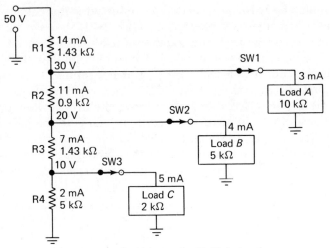

FIGURE 2-22. Voltages for Multiple Loads

properly and at the same time. Working with the given values of current, voltage, and resistance, we can easily see that each circuit does indeed have the proper voltage and current. We have these values because all three loads are functioning normally.

Now let us see what happens when we open switch SW2 and remove the voltage from load B. But first, please observe that, with all switches closed, we have a total resistance of 3.57 kΩ. With 50 V applied to 3.57 kΩ our total current from the voltage source is

$$I_t = \frac{E_a}{R_t}$$

$$= \frac{50 \text{ V}}{3.57 \text{ k}\Omega}$$

$$= 14 \text{ mA}$$

When we open SW2, we cause the total resistance to increase and the total current to decrease. Removing the 5-kΩ load increases R_t to 4.16 kΩ and reduces I_t to 12 mA. Twelve mA through R1 causes a voltage drop of

$$ER1 = IR$$
$$= 12 \text{ mA} \times 1.43 \text{ k}\Omega$$
$$= 17.2 \text{ V}$$

Subtracting 17.2 V from 50 V leaves us 32.8 V across load A

instead of the desired 30 V. The current through load A now rises to 3.3 mA. The current through R2 and R3 is now 12 mA −3.3 mA = 8.7 mA. The combined voltage drop across R2 and R3 is

$$8.7 \text{ mA} \times 2.33 \text{ k}\Omega = 20.3 \text{ V}$$

Then 32.8 V − 20.3 V = 12.5 V. This 12.5 V now appears across R4 in parallel with load C, which is 2.5 V above our desired value of 10 V. This increase in voltage causes the current through load C to increase from 5 mA to

$$I = \frac{E}{R}$$
$$= \frac{12.5 \text{ V}}{2 \text{ k}\Omega}$$
$$= 6.25 \text{ mA}$$

The previous observations reveal that each load on a multiple loaded voltage divider interacts with all other loads. The switch that we opened caused a negligible effect on load A, but it caused current and voltage of load C to increase more than 20 percent. Both of the remaining loads will function reasonably well with these changes, but they will not be operating at peak efficiency. In most ordinary circuits, changes up to 20 percent, or a little more, can be tolerated. Therefore, we find multiple loaded voltage dividers in nearly all solid state power supplies. Without the voltage divider, we would need a separate rectifier circuit for each different level of voltage that we use for the various loads.

VOLTAGE REGULATORS

Voltage from most rectifiers, even after optimum filtering, is expected to vary its level from time to time. In most circuits, these variations can be tolerated. However, many of our other solid state circuits must have precise values of voltages. For these critical circuits, we must limit the voltage variations within very close tolerances. Voltage regulators provide the necessary control on these critical voltage values.

Zener Shunt Regulator A very effective voltage regulator may consist of nothing more than a Zener diode connected in parallel with the load. In effect, this gives us an automatically variable resistor that will adjust its current to compensate for any change in either load current or source voltage. In addition to acting as a variable resistor, the Zener is a highly sensitive device for detecting changes. It detects and corrects all fluctuations before they can alter the voltage across the load. Such a shunt regulator is illustrated in Fig. 2-23.

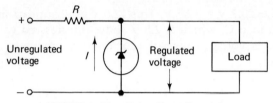

FIGURE 2-23. Zener Shunt Regulator

The series resistor R represents all the resistance between the unregulated voltage source and our Zener diode. It may be a physical resistor or it may not. The purpose of this series resistor is to provide a voltage control when the unregulated input changes. The load current and the Zener current both pass through R. The Zener is selected with a characteristic breakdown equal to the desired value of regulated voltage. The unregulated voltage value minus the voltage drop across R leaves enough voltage to operate the Zener diode in the center of the Zener region. The operating point is point A on the characteristic curve in Fig. 2-24. The zone of regulation then extends between points B and C. This, of course, is reverse current through the Zener; it has a positive potential on the cathode and a negative potential on the anode.

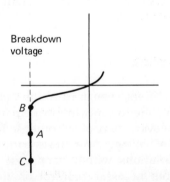

FIGURE 2-24. Zener Control Region

When the load current increases, the Zener feels a slight decrease in its reverse bias and immediately decreases its conduction. The decrease in Zener current exactly matches the increase in load current. The total current remains constant, and the voltage drop across R does not change. We have exactly the opposite action when the load current decreases. The Zener current increases to counteract the change in load current.

We also have fluctuations in the unregulated input voltage that must be countered by our regulator. When the applied voltage increases, it increases reverse bias on the Zener. The Zener reacts with an increase in current which increases the current through R. The additional voltage drop across R cancels the increase in applied voltage to hold the regulated voltage at a constant level. A decrease in the applied voltage triggers an opposite reaction. The Zener current decreases to reduce the voltage drop across R, and this cancels the decrease in the applied voltage.

These Zener diodes are also sensitive to temperature changes. If we expect best results from a Zener shunt regulator, we must make sure that it operates at a constant temperature or has compensation for temperature change. We can do this by placing thermisters or common junction diodes in series with the Zener diode. One method of temperature compensation is illustrated in Fig. 2–25.

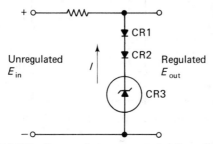

FIGURE 2–25. Temperature-compensated Zener Regulator

The common diodes CR1 and CR2 are forward-biased, and they have a negative temperature coefficient. The Zener diode has a positive temperature coefficient, and it is reverse-biased. The current through this branch has very little opposition, and this opposition will hold constant over a wide range of temperature change. Two junction diodes are used because each of them has a negative temperature coefficient equal to half the positive temperature coefficient of the Zener diode.

An increase in temperature causes the combined resistance of CR1 and CR2 to decrease by the same amount that it causes the resistance of CR3 to increase. The overall result of this combination is a constant value of regulated E_{out} despite all reasonable changes in E_{in}, load current, and temperature.

Transistor Shunt Regulator Circuit components have a tendency to slight characteristic changes because of operating conditions and age. After a period of time, a given circuit may draw a little more or a little less current. A rectifier circuit may deliver a little more or a little less voltage. These characteristic changes, coupled with day-to-day differences in operating conditions, may cause the level of our regulated voltage to be slightly out of tolerance. It may be still regulated, but the level may be a few volts too high or a few volts too low for our purposes. This change in regulated voltage level may be a very critical factor. If so, we need a means of adjusting the level of the regulator output. An adjustable shunt regulator can be constructed by adding a transistor and a parallel resistance branch to our Zener regulator. This type of regulator is illustrated in Fig. 2–26.

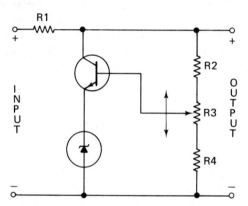

FIGURE 2–26. Transistor Shunt Regulator

The parallel resistor network composed of R2, R3, and R4 has a very high resistance. This high resistance limits the current and diminishes the loading effect of the resistor branch. The potentiometer, R3, can be set to tap off the desired level of voltage on the variable contact. The control for R3 may be either a control knob or a screwdriver adjustment.

The Zener diode controls the potential on the emitter of the transistor. This emitter potential, along with the voltage from R3, determines the bias across the input junction of our transistor. This bias controls the current through the transistor.

An increase in the level of input voltage, or a decrease in output load current, causes an increase in voltage across both parallel branches. This gives us more voltage on the base of the transistor, which increases the forward bias and causes an increase in current through the Zener and the transistor. The increase in current causes an increased voltage drop across R1, which cancels the original action and holds the regulated output constant.

When we have a decrease in the input level or an increase in output load current, we have a decrease in potential across both parallel branches. This gives us less voltage on the base of our transistor, which decreases the forward bias and causes a decrease in current through the Zener and the transistor. This decrease in current results in less voltage being dropped across R1. This action cancels the original action and holds the regulated output constant.

When the regulated output level is out of tolerance, either too high or too low, we can adjust it by varying the setting of R3. When we move the center arm of R3 upward, it taps off more voltage, increases forward bias on the transistor, increases Zener and transistor current, increases the voltage drop across R1, and lowers the level of our regulated output voltage. When we move the center arm of R3 downward, it taps off less voltage, decreases forward bias on our transistor, decreases Zener and transistor current, decreases the voltage drop across R1, and raises the level of our regulated output voltage.

Electronic Series Regulator Shunt regulators are not suitable when we draw large quantities of current or when our load current varies over a certain margin. The series regulators are reliable over wider ranges, can handle more current, and are generally more efficient than the shunt regulators. A series regulator is basically an automatic, variable resistor in series with the load resistance. Our electronic series regulator uses a transistor as the variable resistor. A simple electronic regulator circuit is illustrated in Fig. 2-27.

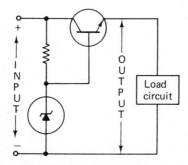

FIGURE 2-27. Electronic Series Regulator

Our series transistor is an amplifier that is very sensitive to minor changes. It makes a very effective automatic, variable resistor. The potential on the base of our transistor is determined by the operating voltage of the Zener diode. This is a relatively high positive potential. The positive potential on the base is higher than the regulated positive potential on the emitter. These two voltages provide forward bias for the transistor. The base potential is held at a constant value, but any change in either input potential or load current will vary the emitter potential and alter the forward bias.

When we have either an increase in input voltage or a decrease in load current, the emitter becomes more positive. As the emitter goes positive, forward bias decreases, transistor resistance increases, voltage drop across the transistor increases, and this cancels the original action. The level of the regulated output is corrected before it has had a chance to make a significant change.

When our input voltage decreases, or the load current increases, we have the opposite reaction. The emitter potential goes in a negative direction, forward bias increases, transistor resistance decreases, voltage drop across the transistor decreases, and this cancels the original action. Again, the regulated output is corrected immediately.

Improved Electronic Series Regulator We can greatly improve our electronic series regulator by adding a few more components. Some of the improvements are greater stability over a wider range of current and voltage, closer tolerances in the regulated output, and a means of adjusting the output level. The circuit in Fig. 2–28 illustrates one arrangement for such a regulator.

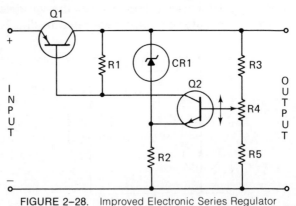

FIGURE 2–28. Improved Electronic Series Regulator

The potential at the emitter of Q2 varies with the Zener current through R2. The base potential of Q2 is determined by the voltage on the movable contact of R4. These two voltages compose the forward bias for Q2 and determine its level of conduction. The voltage drop across Q2 controls the bias and the conduction of Q1, which is our variable, series resistance. The resistance of Q1 is controlled by its level of conduction, and this resistance determines the voltage drop across Q1.

When our input voltage increases, or our output load current decreases, current through CR1 increases and causes an increased positive potential on the emitter of Q2. The base of Q2 moves slightly positive but not as much as the emitter. Our forward bias on Q2 decreases and reduces the collector current. The voltage drop across Q2 increases, and the voltage drop across R1 decreases. This combined action raises the positive potential on the base of Q1 and reduces its forward bias. (This is opposite to Q2 because Q1 is a PNP transistor.) Conduction of Q1 is decreased with a resultant rise in resistance. This action causes the voltage drop across Q1 to increase and cancel the original action.

When our input voltage decreases, or our output load current increases, a reverse action occurs. Zener current decreases, forward bias on Q2 increases, Q2 current increases, voltage across R1 increases, and forward bias on Q1 increases. Q1 now conducts harder and has a reduced resistance. We have less voltage dropped across Q1, which cancels the original action to hold the output at a constant level.

We may have either a control knob or a screwdriver adjustment on the potentiometer, R4. When we move the arm of R4 upward, we reduce the forward bias on Q2. This reduces the conduction of Q2 and Q1 and reduces the level of our regulated output voltage. When our regulated output is too low, we correct it by moving the arm of R4 downward. This action increases the conduction of Q2 and Q1 to raise the level of our regulated output voltage.

CONVERTERS

We have already studied ac to dc converters. A transformer combined with a rectifier is such a converter. It converts alternating voltage and current to direct voltage and current. But there are other converters that play an important role in our solid state power supplies. We have dc to dc converters which raise the level of direct voltage. Also we sometimes use dc to ac converters to change direct voltage back to alternating voltage.

DC to AC Converter One of our main objectives in converting dc back to ac is to provide voltage for operating various types

60 Energy Conversion Circuits

of ac motors. There are mechanical devices for this type of conversion, but solid state circuits are lighter in weight, and they are generally more efficient than vibrators and rotary inverters. An electronic dc to ac converter normally uses a free-running oscillator to break the direct voltage into square waves, and then the square waves are passed through a transformer. The basic principles of this action are illustrated in Fig. 2–29.

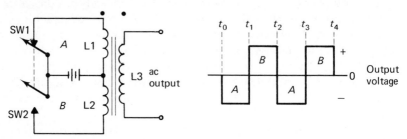

FIGURE 2–29. Switching Circuit and Resulting Waveforms

In this simplified circuit, we can easily follow the switching action. Switches SW1 and SW2 are ganged together so that as one opens the other closes. In the illustration, SW1 is closed and circuit A is activated. Current from the battery passes through SW1, down through L1, and back to the battery. While SW1 is closed, waveshape A is developed in circuit A and coupled to L3. At time $t1$, SW1 opens and SW2 closes. Electrons now flow upward through L2 as the waveshape changes from A to B. Polarities reverse each time we reverse the switch positions. Many motors can utilize this type of square wave almost as well as a sine wave, so all we really need is an automatic switch, and transistors make good automatic switches. When we replace the switches with transistors, our circuit becomes as illustrated in Fig. 2–30.

We now have a free-running oscillator circuit that functions in the same way as the switching arrangement previously discussed. Transistors Q1 and Q2 are identical, PNP, alloy junction, power transistors. Transformer windings L2 and L3 are collector coils that represent the main primary windings and the load coils for the transistors. Coils L1 and L4 are feedback windings. L5 is the output secondary, and it usually has more turns than either L2 or L3 in order to provide a step-up in the voltage. Resistors R1 and R2 form a voltage divider to provide forward bias for both input junctions. The transistors are now serving as both amplifiers and high-speed switches.

The conducting transistor provides feedback to enhance its own conduction while holding the other transistor cut off. This

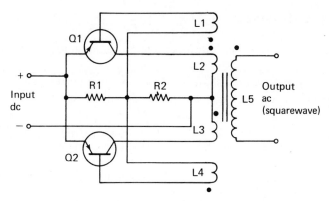

FIGURE 2-30. DC to AC Converter

regenerative action continues until the conducting transistor approaches saturation. The feedback then weakens, which allows the quiescent transistor to start conducting and take over the circuit. Each transistor conducts half the time and is cut off half the time. As one transistor switches on, the other switches off.

The waveshapes on L2 and L3 are very nearly square waves. These square waves are coupled across the transformer and stepped up in amplitude, and become the output alternating voltage. The dots on the coils indicate points of like polarity.

Many types of motors can be driven by square waves. The motor may act as a filter and react only to the fundamental frequency that composes the square wave. The peak value of this fundamental frequency is $4/\pi$ times the square wave voltage. Therefore, our square wave voltage must equal the rated sine wave voltage times the factor

$$\frac{\pi}{4 \times 0.707}$$

This means that a motor that is rated at 115 V ac would require about 129 V ac if it were operated by a square wave voltage.

DC to DC Converter The dc to dc converter first converts dc to ac and then changes the ac back to dc. So if we connect a suitable rectifier circuit to the secondary coil in Fig. 2-30, our dc to ac converter becomes a dc to dc converter. The rectifier may be half wave, full wave, bridge, or whatever best provides the desired output. We can even use voltage multipliers if we want a very large value of voltage.

REVIEW EXERCISES

1. Why do power transformers used in solid state power supplies have to be especially designed?
2. All power frequences fall into what frequency range?
3. A particular power transformer is designed to handle what portion of the power frequencies?
4. A particular transformer has an input power of 200 watts and is delivering 190 watts to the output load. What is its percent of efficiency?
5. What happened to the 10-watt difference between input power and output power in item 4?
6. Name three kinds of transformer energy losses and state the cause of each.
7. What is the primary to secondary turns ratio of a transformer that has 120 V on the primary and 10 V on the secondary?
8. The transformer in item 7 has 96 mA in the secondary. What is the value of current in the primary?
9. What is indicated by the PIV rating of a diode?
10. In a rectifier circuit what is the advantage of connecting two or more diodes in
    a. Series?
    b. Parallel?
11. What type of power requirements are best suited by
    a. Half-wave rectifiers?
    b. Full-wave rectifiers?
12. What is the purpose of a rectifier?
13. With respect to the input frequency, what is the ripple frequency of
    a. Half-wave rectifiers?
    b. Full-wave rectifiers?
14. Draw a 2:1 step-down transformer, with phase inversion, feeding a half-wave, positive rectifier. Show two input cycles and the corresponding output.
15. If we assume that the transformer in item 14 has a peak-to-peak input of 120 V, what is the peak-to-peak value of the secondary voltage, and what is the maximum value of the output voltage from the rectifier?
16. What are the relative values of the peak voltage on a transformer secondary and the maximum output from a full-wave rectifier connected to this transformer? Explain why this relationship exists.

17. How many wires are normally used to distribute a three-phase voltage? Under what conditions may the ground wire be eliminated?
18. What is the purpose of a power supply filter?
19. Under what conditions can adequate filtering be performed by a single capacitor in parallel with the load resistor?
20. What is the function of a choke coil in a filter? What condition makes the coil essential?
21. What type of filter is best suited for loads that require large quantities of current?
22. What are the characteristics of a load circuit which enables it to be powered by a voltage multiplier without further filtering?
23. Draw a transformer with a simple voltage doubler, indicating charge paths, discharge path, and polarity of the output voltage.
24. Suppose that the transformer in item 23 is a 1:5 step-up with 50 V ac on the primary. What is the amplitude of the voltage across the load resistor? Explain your answer.
25. Figure 2–31 is a cascade voltage doubler. The peak voltage on the secondary of the transformer is 110 V. Indicate on the diagram the charge path for both capacitors, the quantity and polarity of the voltage on each capacitor, and the level and polarity of the output voltage.

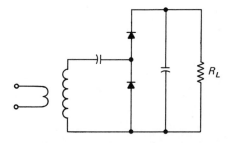

FIGURE 2–31. Cascade Voltage Doubler

26. What is the name of the device that enables us to take several values of voltage from a single voltage source?
27. State the purpose and describe the distinguishing features of a
    a. Rheostat.
    b. Potentiometer.
28. What are the two primary variations that are compensated for by a voltage regulator?
29. What is the primary difference between a shunt regulator and a series regulator?

30. What type of regulator is best suited for regulating voltage to a load that requires medium voltage with large variations in current?
31. What advantage is gained by connecting two junction diodes in series with a Zener shunt diode in a shunt regulator? Why is this possible?
32. What are the advantages of the regulator in Fig. 2-28 over that in Fig. 2-27?
33. Draw a switching circuit to illustrate the action of a dc to ac converter.
34. Draw the output waveshape of a dc to ac converter.
35. Describe the difference between a dc to ac converter and a dc to dc converter.

Chapter Three

Principles of Amplification

The pressing need for a better amplifying device was the force behind the invention of the transistor. Very few practical electronic circuits can be constructed without creating the need for an amplifier. Amplification is the process that increases the magnitude of a signal. We have voltage amplifiers, current amplifiers, and power amplifiers to increase the individual signal components as needed. In this chapter, we review some of the important characteristics of transistors. The common circuit arrangements are analyzed, along with biasing circuits and stabilization techniques.

BASIC AMPLIFIERS

We begin by analyzing amplifier circuits in a general way. This will enable us to concentrate on general amplification principles that are common to all types of amplifiers. One of the primary objectives of studying basic amplifiers is to develop an understanding of how a transistor can increase the amplitude of a signal.

Amplification Features Normally we use direct voltage to bias a transistor and use the transistor to amplify an ac signal. We consider the bipolar transistor as a current-sensitive device because small changes in bias voltage cause significant changes in the current through the transistor. The current existing as a result of bias voltages alone are dc. When we introduce an ac signal into the input circuit, it alters the input bias and causes a small change of the input current. This change is greatly magnified through the transistor, and we have a large change in the output current. Of course, we also have voltage and power amplifiers, as well as the current amplifier just described. Figure 3–1 is a schematic diagram of a basic amplifier.

66 Principles of Amplification

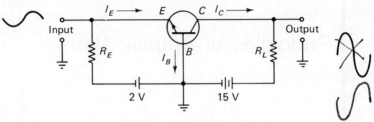

FIGURE 3–1. Basic Amplifier

Let us suppose that the load resistor R_L is 10 kΩ and the static I_C is 0.5 mA. Further, suppose the input signal to have a peak amplitude of one volt. This gives us a 5-V drop across R_L and leaves a collector voltage of 10 V.

The change in bias, caused by the input signal, swings from 2 V to 3 V and back to 2 V with the negative alternation of the signal. On the positive alternation, bias swings from 2 V to 1 V and back to 2 V. This change in bias causes a small but significant change in the input current. Let us say that this input current changes from 0.5 mA to 0 mA, back to 0.5 mA and up to 1 mA. These are not good operating parameters, but they will serve for illustration. Most (98 percent) of this input current change is passed on to the collector. As the collector current drops toward 0 mA, the voltage drop across R_L decreases and allows the output voltage to go more positive. When I_C reaches zero, the output voltage is the same as the collector-base bias, 15 V in this case. As our input signal swings negative, the input forward bias increases and causes the emitter current to move toward our assumed value of 1 mA. As I_C rises from zero to 1 mA, the voltage drop across R_L increases from zero to 10 V. At this same time, the output voltage drops from 15 V to 5 V. So a 1-V signal causes a change of 10 V in our output voltage: a gain of 10 in the voltage amplification.

The transistor symbolized in Fig. 3–1 is an NPN. The symbol for the PNP is the same, except that the arrow on the emitter points to the base. The circuit for a PNP is also the same, except that we must reverse the polarity of both batteries. In fact, we can change any transistor circuit from an NPN to a PNP (or vice versa) by simply reversing the polarities of all biasing voltages and exchanging the transistors.

Amplifier Nomenclature   We identify amplifiers in several different ways: by the frequencies they amplify, the portion of the signal they amplify, the element of the signal they concentrate on, the circuit configuration, and several other categories. We have audio frequency amplifiers for audio frequencies, radio frequency amplifiers for radio signals, and video frequency amplifiers for video signals. We have amplifiers classed as A, B, AB, and C to specify the percentage of each input cycle that the amplifier will conduct. We have current amplifiers to amplify current, voltage amplifiers to amplify voltage, and power amplifiers to amplify power. The transistor may be found in one of three circuit configurations: the common base, the common emitter, or the common collector. The circuit configuration is chosen that will best serve the specific purpose for which the amplifier is designed.

Classes of Amplifiers Amplifiers are classed according to the percentage of the input cycle that the transistor is conducting. The transistor is our amplifying vehicle, and the entire stage is the amplifier circuit, which we normally refer to simply as the amplifier.

In a class A amplifier, we bias the transistor so that it conducts all the time. We never allow this amplifier to be overdriven. Therefore, the entire input signal is reproduced in the output.

With the class B amplifier, we bias it so that the transistor will be cut off about 50 percent of the time. If we pass a sine wave through a class B amplifier, one alternation will pass and the other will be eliminated from the output.

We bias the class AB amplifier so that it will cut off during each cycle of the input signal but will remain cut off less than 50 percent of the time. This amplifier eliminates a portion of one alternation from the output.

The bias on a class C amplifier will cause it to be cut off more than 50 percent of the time. This amplifier will eliminate a portion of both alternations from the output. Figure 3–2 illustrates the results that we may expect when we pass a sine wave through an amplifier of each class.

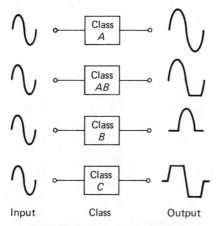

FIGURE 3–2. Amplifier Classes

68 Principles of Amplification

The class of amplifier used in a given situation is dictated by the signal to be amplified and the desired shape of the output. In the case of a high fidelity audio amplifier, only the class A will do a proper job. In several wave-shaping situations, the class C amplifier fits the job precisely.

Circuit Configurations   There are many ways to arrange the components in an amplifier circuit, but all of these ways fit into three categories. These are common base, common emitter, and common collector. We derive these three names from the transistor element, which is common to both the input and output sections of the circuit. We sometimes find the word "ground" used instead of "common." We shall now examine each of these configurations in detail.

COMMON BASE

In the common base configuration, the base of the transistor is common to both the input and output circuit. This arrangement is relatively popular, and we frequently find it used for either voltage or power amplification. It will produce a voltage gain up to 1500 and a power gain of as much as 30 dB. We consider that the power level doubles for each increase of 3 dB or halves for each decrease of 3 dB.

Bias for the Common Base   The transistor in Fig. 3–1 is in a common base arrangement. The bias shown in that circuit is symbolical of the common base biasing circuits for an NPN transistor. The same configuration, with reversed polarities, would be symbolical of the biasing arrangement for a PNP transistor. However, it is not necessary that we use two batteries. Figure 3–3 shows an arrangement that provides bias for both input and output junctions with only one battery.

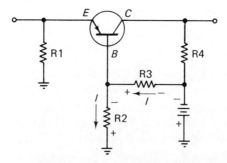

FIGURE 3–3.   Single Battery Bias

In this circuit we have a PNP transistor in a common base configuration, and the single battery is providing all the bias. R2 and R3 form a voltage divider. The current through these resistors, direction as indicated, will hold the base positive with respect to the collector and negative with respect to the emitter. This provides our input junction with forward bias and our output circuit with reverse bias.

The direct current through the transistor (I_c) is in addition to the voltage divider current just discussed. The total transistor current leaves the negative terminal of the battery and splits at the junction of R3 and R4. About two percent of this total is our base current. It passes through R3 and enters the base of the transistor. The remaining 98 percent is our collector current. It goes through R4 and enters the transistor collector. The current recombines in the transistor base and emerges as our emitter current. It passes through R1 to ground, and since the positive terminal of the battery is grounded, the path has been completed.

With the exception of portable devices, we seldom find batteries in electronic equipment. That is why the battery bias circuits were referred to as symbolical. Generally we find a power supply that changes the alternating voltage to direct voltage, and the direct voltage may be regulated to a very close tolerance. Voltage of this type is suitable for bias voltage, and it eliminates the problem of maintaining batteries. The direct voltage is applied across a voltage divider from which various quantities of voltage can be tapped off as needed. Figure 3–4 illustrates a biasing circuit that uses this type of voltage divider bias.

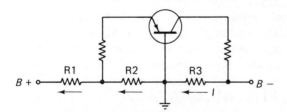

FIGURE 3–4. Voltage Divider Bias

Resistors R1, R2, and R3 are parts of a voltage divider, and they are usually located with our power supply. The current from B− to B+ provides voltage drops that cause the base to be more positive than the collector and more negative than the emitter. Since our transistor is a PNP, these are the polarities that we require for proper bias.

The voltage across the voltage divider is a regulated voltage

70 Principles of Amplification

in most cases. Since the transistor operates with ac, precautions must be taken to prevent the ac from mixing with the dc. This is accomplished by using ac bypass capacitors along the voltage divider. These capacitors are connected from the various taps to ground. The capacitors are opens to the dc and shorts to the ac. This enables us to have a regulated direct voltage biasing circuit with several points physically connected to lines that carry ac.

On most schematics, we find the transistor in one location and the power supply in another. The points of connection are identified by code, and the connecting lines are eliminated. This is illustrated in Fig. 3-5.

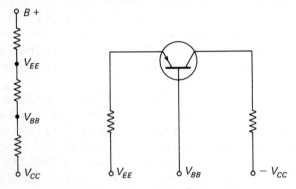

FIGURE 3-5. Bias Voltage Designators

Notice that specific points on the voltage divider are labeled. V_{EE} designates emitter supply voltage; V_{BB} is base supply voltage; V_{CC} is the supply voltage for the collector. You will notice the same designations on the transistor circuit. When these values are negative, the designator is preceded by a minus sign, as shown with the V_{CC} in the diagram. It is common practice to identify direct voltages and currents with capital letters and to use lowercase letters for alternating voltages and currents.

Of course, the biasing arrangements discussed here, in connection with the common base amplifier, are also true for most other transistor applications.

Voltage-current Characteristics We may obtain transistor data charts or specification sheets from our dealer. These charts are prepared by the manufacturer, and they list many of the static characteristics for a particular transistor. We usually find a graph similar to that in Fig. 3-6 as part of this data.

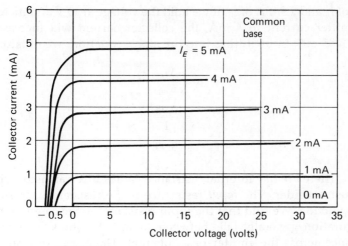

FIGURE 3-6. Common Base Characteristic Curves

This is a family of voltage-current characteristic curves for a particular transistor when it is connected as a common base amplifier. It shows the emitter current and the collector current for a wide range of collector voltages. As you can see, the emitter current lines are almost parallel to the collector current lines. Notice that for any given value of collector voltage, our emitter current is only slightly higher than the collector current. This is in keeping with our earlier examples, where we had collector current representing 98 percent of the total current.

We have a name for this ratio of collector current to emitter current; we call it alpha (α). It is the current amplification factor for a common base amplifier. It may vary from 0.95 to 0.99 with different transistors, but it is always less than unity. Alpha is calculated by

$$\alpha = \frac{I_C}{I_E}$$

We are actually dealing with a dc ratio here, but we want to adapt it to ac amplification. For this reason, the alpha equation is frequently expressed as a change ratio at some fixed value of collector voltage, like this:

$$\alpha = \frac{\Delta I_C}{\Delta I_E}$$

For instance, on the 15-V line of the graph, suppose that we have

3 mA of I_E and 2.9 mA of I_C. If we hold the voltage at 15 V and increase the emitter current to 4 mA, the collector current will increase to about 3.85 mA. Substituting into the alpha equation, we obtain:

$$\alpha = \frac{\Delta I_C}{\Delta I_E}$$

$$= \frac{0.95}{1}$$

$$= 0.95$$

There are *two* internal resistance values in a transistor which we must consider. The *input* resistance is the resistance across the input junction. We find that the input resistance of the common base configuration generally falls between 30 and 150 ohms. This is the resistance across the emitter-base junction. The *output* resistance is the resistance across the output junction. In the common base configuration, this is between 300 and 500 kilohms. This is the resistance across the collector-base junction. These resistive values are measured with the external leads open.

Voltage Gain Voltage gain (A_v) is the ratio of output voltage to input voltage. We normally find that this ratio is between 300 and 1500 in our common base amplifier. Since the input voltage is V_e and the output voltage is V_c, we may calculate voltage gain as follows:

$$A_v = \frac{V_c}{V_e}$$

Again we are dealing with alternating values that may be interpreted as variations of the direct bias voltage. So you may see the gain formula expressed as

$$A_v = \frac{\Delta V_c}{\Delta V_e}$$

Another means of calculating voltage gain is to multiply the resistance ratio by the current gain. We have already established alpha as the current gain. The resistance value of R_L is the output resistance, and the resistance of the emitter-base junction is the input resistance. The calculation is performed in this manner:

$$A_v = \alpha \left(\frac{R_{\text{out}}}{R_{\text{in}}} \right)$$

Common Base

Power Gain The common base amplifier does a reasonably good job as a power amplifier. It produces a power gain between 20 and 30 decibels. We symbolize power gain as G. Power gain is the ratio of output power to input power. The input power is the emitter power. The output power is the collector power. We calculate power gain as follows:

$$G = \frac{P_c}{P_e}$$

Of course, we may express power in other terms. Power gain is equivalent to the product of current gain times voltage gain. Power gain is also equivalent to the square of the current gain times the resistance ratio. We can express both of these ideas as follows:

$$G = \alpha^2 \left(\frac{R_{out}}{R_{in}} \right)$$

Signal Phase Relationship We have no phase shift through a common base amplifier. This can be verified by examining the phase of the input and output voltage in Fig. 3-7.

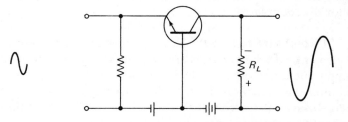

FIGURE 3-7. Signal Phase Relationship

In this NPN circuit, the forward bias holds the emitter negative with respect to the base. As the input signal swings negative, it adds to the bias voltage and increases the forward bias. This gives us an increase in current in all three elements of the transistor. The increase in collector current causes an increase of voltage drop across R_L with a polarity as shown. This develops a negative swing in the output voltage. When our input signal swings positive, forward bias is reduced, current is reduced, and output voltage swings positive.

Cutoff Frequency The diffusion capacitance inside a transistor effectively places a capacitor in parallel with each junction resistor. At low frequencies, the capacitor acts as an open and has no effect on the gain of the amplifier. As we increase the signal frequency, the capacitive reactance decreases. There comes a point at which the capacitor is a short across the junction, and the transistor has no gain.

The cutoff frequency is the frequency that reduces the current gain to 0.707 of its low frequency value. We find this on data charts as $f\alpha$. If we have a transistor with an alpha of 0.98, our $f\alpha$ is the frequency that gives us 0.707×0.98. In other words, our cutoff frequency produces an alpha of 0.693.

Characteristic Summary

1. Input resistance: 30 to 150 ohms.
2. Output resistance: 300 to 500 kilohms.
3. Voltage gain: 300 to 1500.
4. Current gain: less than unity.
5. Power gain: 20 to 30 dB.
6. No phase shift.

COMMON EMITTER

The common emitter is the most popular of our three configurations. Let us take a close look and see if we can discover some of the reasons for this popularity.

Circuit Arrangement Now we have the emitter as the common element, and the circuit is arranged similarly to that in Fig. 3–8. We apply the input signal to the base, but it is still developed across the emitter-base junction. This is our input junction, and it must be forward-biased. We accomplish this with an NPN by making the base positive with respect to the emitter.

Our output is taken between the collector and emitter termi-

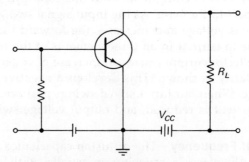

FIGURE 3–8. Common Emitter Amplifier

nals, across R_L. The collector of this NPN transistor must be positive with respect to both emitter and base. V_{CC} accomplishes this when connected as indicated. Single-battery bias, or voltage divider bias, may be arranged in a manner similar to that discussed under the common base amplifier.

Voltage-current Characteristics A family of voltage-current characteristic curves is illustrated in Fig. 3–9. These data are also supplied by the manufacturer. This family of curves applies to a particular transistor when it is connected as a common emitter amplifier.

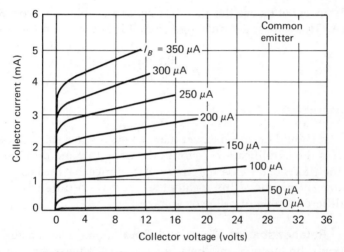

FIGURE 3–9. Common Emitter Characteristic Curves

Our important ratio now is collector current to base current. We call this ratio beta β, and it is the amplification factor for the common emitter amplifier. For any given value of collector voltage, we calculate beta as follows:

$$\beta = \frac{I_C}{I_B}$$

For instance, on the 12-V collector voltage line, suppose that we have an I_B of 350 μA. This gives us an I_C of 5 mA, and

$$\beta = \frac{5 \text{ mA}}{350 \text{ }\mu\text{A}}$$
$$= 14.28$$

Please notice that this is a dc ratio. When it appears on data sheets it is referred to as dc beta. But we are more interested in the ac ratio, which we call ac beta.

When calculating ac beta, we still hold to a constant value of collector voltage. Now we change the base current by some specified amount and note the resulting change in the collector current. Then ac beta is calculated as follows:

$$\beta = \frac{\Delta I_C}{\Delta I_B}$$

For instance: Let us use the 12-V line again and change I_B from 300 to 350 μA. This causes I_C to change from 4.2 to 5 mA, and ac beta is

$$\beta = \frac{0.8 \text{ mA}}{50 \ \mu\text{A}}$$
$$= 16$$

It should be emphasized that ac beta is still a static characteristic. It is the amplification factor of the transistor and does not consider the effects of the external circuit. Also notice that the base current lines *are not* parallel. This means that we have a different beta for each different value of collector voltage.

Relationship of Alpha and Beta Some manufacturers *do not* supply both alpha and beta. This makes it necessary for us to convert from one to the other. There is a direct relationship that makes this conversion a simple operation. The relationship between alpha and dc beta is expressed as

$$\beta = \frac{\alpha}{1 - \alpha}$$

or

$$\alpha = \frac{\beta}{1 + \beta}$$

Suppose that we have an alpha of 0.98 on a data sheet, but wish to use the transistor in a common emitter configuration. We substitute for alpha into the beta equation and solve in the following manner:

$$\beta = \frac{\alpha}{1-\alpha}$$

$$= \frac{0.98}{1-0.98}$$

$$= \frac{0.98}{0.2}$$

$$= 49$$

The value of beta tells us what we may expect as maximum current amplification. This value may range from 20 to 200 with various transistors. Suppose that our data sheet shows a beta of 180, but we want to use the transistor as a common base amplifier. Now we must convert beta to alpha. We substitute beta into the alpha equation, and solve as follows:

$$\alpha = \frac{\beta}{1+\beta}$$

$$= \frac{180}{1+180}$$

$$= \frac{180}{181}$$

$$= 0.9944$$

Current Gain   Remember that beta is the amplification factor only. It *does not* tell us the actual amplification when we place the transistor in a circuit. On paper the current gain appears much the same as beta:

$$A_i = \frac{I_c}{I_b}$$

The values are now dynamic, and both collector and base currents are affected by the external circuit. We should expect the actual current gain to be somewhat less than beta. The normal current gain falls between 25 and 60 for a common emitter amplifier.

Voltage Gain   The actual voltage gain is the product of current gain times the ratio of output load resistance to the input resistance. Until we determine the value of the actual current gain, beta may be used to represent our current gain. Therefore, voltage gain is approximated by this equation:

78 Principles of Amplification

$$A_v = \beta\left(\frac{R_L}{R_{in}}\right)$$

Input resistance generally falls between 500 and 1500 ohms, and the load resistance is usually considerably less than the transistor's output resistance. Since the output resistance of the common emitter falls between 30,000 and 50,000 ohms, our load resistor should be between 10,000 and 20,000 ohms. Assuming that our transistor has an input resistance of 500 ohms and a beta of 50, and that we use a load resistance of 10,000 ohms, we can calculate the voltage gain as follows:

$$A_v = \beta\left(\frac{R_L}{R_{in}}\right)$$
$$= 50\left(\frac{10\ k\Omega}{500\ \Omega}\right)$$
$$= 50(20)$$
$$= 1000$$

The values in the previous example reflect maximum expectations for voltage amplification in the common emitter. Normal voltage gain for this configuration is 300 to 1000.

Signal Phase Relationship   The common emitter amplifier produces a 180-degree phase shift between the input and output signals. Let us examine the schematic in Fig. 3–10 to determine why this phase shift takes place.

This is a common emitter amplifier with single battery bias. As the battery is placed, our emitter is negative with respect to both

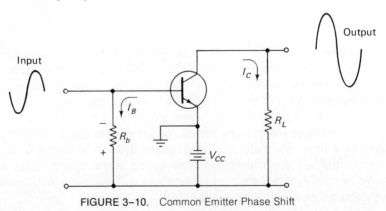

FIGURE 3–10.   Common Emitter Phase Shift

base and collector. Since we have an NPN transistor, these potentials place forward bias on the input and reverse bias on the output. The base potential is V_{CC} less the voltage drop across R_b. This leaves the base only slightly more positive than the emitter. The potential at the collector is V_{CC} less the voltage drop across R_L.

When we apply an ac signal to the base, it will ride on the bias potential. This means that the bias increases and decreases at an ac rate. The negative alternation of the signal decreases the forward bias and causes the base current to decrease. A small decrease in base current results in a large decrease in collector current. This decreases the voltage drop across R_L, and the collector potential rises toward V_{CC}.

The positive alternation of the signal increases the forward bias and causes an increase in base current. A small increase in base current results in a large increase in collector current. This increases the voltage drop across R_L, and the collector potential decreases. This action constitutes a 180-degree phase inversion of the input to output signal.

Power Gain Power gain, as always, is the ratio of output power to input power. The normal power gain of the common emitter ranges from 25 to 40 decibels. We can calculate this gain by multiplying the current gain by the voltage gain; like this:

$$G = A_i A_v$$

Since we must rely on the amplification factor until the actual circuit has been constructed, we can arrive at the approximate value of power gain by using the following equation:

$$G = \beta^2 \left(\frac{R_L}{R_{in}} \right)$$

Summary of Characteristics We should have concluded from this discussion that the common emitter configuration is an excellent amplifier for either current, voltage, or power. Let us summarize the outstanding characteristics.

1. Input resistance: 500 to 1500 ohms.
2. Output resistance: 30,000 to 50,000 ohms.
3. Voltage gain: 300 to 1000.
4. Current gain: 25 to 50.
5. Power gain: 25 to 40 dB.
6. 180-degree phase shift between input and output.

COMMON COLLECTOR

The common collector is not the best choice when our chief concern is amplification. It has no voltage gain, and both common base and common emitter circuits excel it as a power amplifier. It matches the common emitter for current amplification but does not excel it. Then why do we bother with this circuit? The main reason is impedance matching. Its high input and low output impedance enables matching loads of widely differing impedance values. This, coupled with its ability to produce a reasonable gain of either current or power, makes this a very valuable circuit with many applications. We frequently find the common collector referred to as either a grounded collector or an emitter follower.

Circuit Arrangement and Bias Any of the biasing methods we have previously discussed may be applied to this circuit. With an NPN, the voltages must be such that the collector is positive with respect to both base and emitter. Of course, this is reversed when we use a PNP transistor. Figure 3–11 illustrates an NPN used as a common collector amplifier with voltage divider bias.

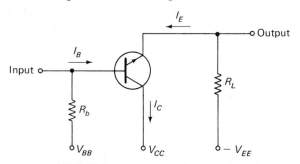

FIGURE 3–11. Common Collector Amplifier

Our common collector amplifier has a very small positive potential for the V_{BB}, and V_{CC} is a relatively large positive value. V_{EE} may be a separate negative source or simply the negative end of V_{CC}. This arrangement places the collector at a potential that is positive with respect to both base and emitter. Since our transistor is an NPN, these potentials give us forward bias across the input junction (collector-base) and reverse bias across the output.

Current Gain Manufacturers *do not* normally furnish separate characteristics for the common collector amplifier. However, we may deduct most of the necessary information from a standard data sheet and a family of characteristic curves for the common emitter. As we previously mentioned, these characteristic graphs for the common emitter are readily available. These graphs are all similar to that shown in Fig. 3–9. They plot collector current I_C against base current I_B for various levels of collector voltage. We have no established symbol for the current gain factor in the common collector. However, the maximum current gain is equivalent to I_e divided by I_b, if a constant value of collector voltage is assumed. Expressing this ratio as an equation, we have

$$A_i = \frac{I_e}{I_b} \quad \text{or} \quad A_i = \frac{\Delta I_E}{\Delta I_B}$$

Looking back at our common emitter graph in Fig. 3–9, we find that I_b is listed, but not I_e. But since I_e is always equal to the sum of $I_c + I_b$ (in any configuration), we can restate the current gain equation as follows:

$$A_i = \frac{I_c + I_b}{I_b}$$

Armed with this information, we can use the data from the common emitter graph and calculate the current gain for a common collector. For example, let us find the current gain of a common collector, using the graph in Fig. 3–9. Our collector voltage is 16 V, and we change the base current from 200 µA to 250 µA. Notice on the graph that doing so causes an I_c change of 0.9 mA. Substituting into our equation, we have

$$\begin{aligned}
A_i &= \frac{I_c + I_b}{I_b} \\
&= \frac{0.9 \text{ mA} + 50 \text{ µA}}{50 \text{ µA}} \\
&= \frac{(0.9 \times 10^{-3}) + (50 \times 10^{-6})}{(50 \times 10^{-6})} \\
&= \frac{950}{50} \\
&= 19
\end{aligned}$$

Of course, we are dealing with a static characteristic, but with this transistor in a common collector configuration, we can expect a maximum current gain of 19. This figure is somewhat low for a common collector. We normally expect a current gain between 25 and 50.

82 Principles of Amplification

Voltage and Phase Relationship We have no phase inversion of signals through our common collector. This configuration also gives us a voltage gain that is less than unity. Let us examine the schematic in Fig. 3–12 to see why this amplifier reacts in this fashion.

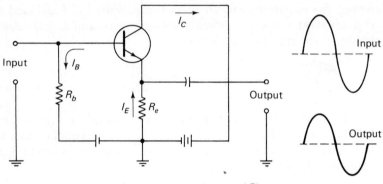

FIGURE 3–12. Voltage and Phase

This diagram shows a properly biased NPN transistor connected as a common collector amplifier. Forward bias gives us a positive potential on the base. As our input signal swings positive, it increases the forward bias and causes an increase in current. The rising current causes the potential on the emitter to increase. By the same token, when our input signal swings negative, it decreases the forward bias on the input junction. This causes a decrease in current and a resulting negative swing of the emitter voltage. So we see that the output across R_e is always in phase with the input that is developed across R_b.

Notice that the output voltage sine wave is slightly less in amplitude than the input sine wave. The reason for this slight loss in voltage amplitude is this:

> The input voltage is equal to the sum of the output voltage plus the voltage dropped across the base-emitter junction.

The voltage drop across this junction is very small, but it is always the difference between the input and the output voltage. Analyze the diagram in Fig. 3–13 for further clarification. Here we see the junction resistance R_J in series with the load resistor. This branch in turn is in parallel with the base resistor R_b. Any change in input voltage appears across both of the parallel branches. Only a portion of this change, ER_L, is tapped off for our output voltage. The voltage gain A_v is V_{out}/V_{in}. Since V_{out} is the voltage across R_e and V_{in} is the

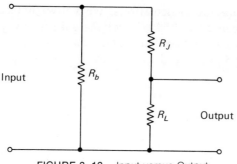

FIGURE 3-13. Input versus Output

voltage across R_b, V_{out} is always slightly less than V_{in}. Therefore, $A_r < 1$.

Resistance The input resistance for a transistor in a common collector circuit is very high. We can expect this resistance value to be between 20 kΩ and 500 kΩ. Because of this high resistance, we require a high input voltage to produce a small input current.

On the other hand, the output resistance of the common collector is very low. It is the resistance of the load resistor or R_e. This value is only a small percentage of the input resistance. Of course, we can change the output resistance within a reasonable range by changing the size of the load resistor. This feature enables us to use the common collector amplifier to match the impedances between high impedance circuits and low impedance circuits. One place where this matching is needed is between the output of an amplifier and the input to a coaxial cable. The common collector amplifier can fill this gap by matching both the impedances, and while doing so, it will give us amplification of both current and power.

Power Gain Since one of the important functions of a common collector is to amplify power, we naturally expect to find an indication of power gain on the data sheets. But on second thought, both current gain and voltage gain affect power gain, and on the other hand, the ohmic value of our emitter resistance affects both current gain and voltage gain. So how do we determine the power gain? The equation we used with the common emitter is also good in any circuit:

$$G = A_i A_r$$

Power gain is always equal to the product of current gain times voltage gain.

84 Principles of Amplification

From previous calculations, we know that current gain is normally between 25 and 50 and voltage gain is always slightly less than unity. Let us assume that a particular amplifier has an $A_i = 42$ and an $A_r = 0.95$ and use these values to calculate power gain. We then have

$$G = A_i A_v$$
$$= 42 \times 0.95$$
$$= 39.9$$

A little thought on this point will reveal that the power gain in a common collector amplifier is always just slightly less than the current gain.

Summary of Characteristics This discussion should have clearly revealed that the common collector amplifier is a valuable circuit that fits very specific needs. We shall find it used less frequently than the common emitter, but it can be used in places where the common emitter cannot function properly. Notably we use the common collector as a current or power amplifier driver in positions where impedances are difficult to match. Here are the outstanding characteristics:

1. Input resistance: 20 to 500 kΩ.
2. Output resistance: 50 to 1000 Ω.
3. Voltage gain: less than unity.
4. Current gain: 25 to 50.
5. Power gain: 10 to 20 dB.
6. No phase inversion between input and output.

BIAS STABILIZATION

So far we have concerned ourselves only with the proper polarity and approximate levels of bias voltage that enable our transistors to function. In a practical circuit there are several other factors to consider when we are establishing our bias. Some of the factors that affect bias are leakage current, barrier potential, changes in circuit conditions, variations in transistor parameters, and circuit stability factor.

Operating Point A transistor's operating point, the set of conditions that enable proper amplification, is determined by our dc operating conditions. The dc bias as previously discussed, along with the circuit configuration and load resistance, sets this operating point. This point gives us a rest value, a no-signal value, for both input and output signal to ride upon. You may consider this rest value and the signal changes to be both current and voltage values. However, when we are dealing with bipolar transistors, most of our references are to current values. Our bias values of both current and voltage are direct values. Our signal values of both current and voltage are the alternating values that swing both ways from the operating point. A properly biased circuit places the operating point at a specified position somewhere between the limits of the amplifying action. These limits are cutoff on the transistor at one extreme and saturation at the other.

Transistor Cutoff We consider a transistor to be cut off when the collector current reaches minimum practical value. This minimum practical value is not quite zero. The output is reverse-biased, and some leakage current will be present. Let us use a common emitter amplifier as a sample circuit for a more specific discussion. Consider the circuit in Fig. 3–14.

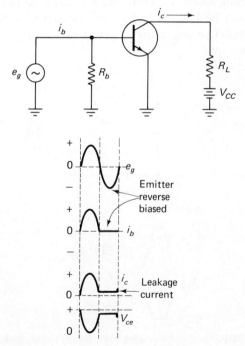

FIGURE 3–14. Common Emitter Driven to Cutoff

86 Principles of Amplification

Here we use the symbol for an ac generator as our signal source. This generator is simply the previous stage, whatever it happens to be, and this previous stage is always considered as our signal source. The designation e_g signifies the signal voltage from this generator. Our dc bias, in this case, has established an operating point near cutoff. As the signal swings positive, we have no problem; this aids the forward bias on the emitter-base junction. When the signal swings negative, our input junction is reverse-biased. When the input bias reaches zero, all current from emitter to base ceases; we consider this as the cutoff point. At cutoff the leakage current is still present from emitter to collector. We call this leakage current I_{ceo}, which simply means: current in the collector when the base is cut off.

This amplifier is operating class B with a resulting clipping or rectification of the output signal. If we wish to avoid this cutoff limiting, we must arrange our bias so that the input signal can never reduce the input junction bias to zero. We must have this junction forward-biased for all values of the input signal.

Transistor Saturation   Another form of limiting or clipping of a signal occurs when we allow our transistor to saturate. Saturation occurs when both junctions of the transistor become forward-biased. We shall use the common emitter circuit in Fig. 3–15 to illustrate this action.

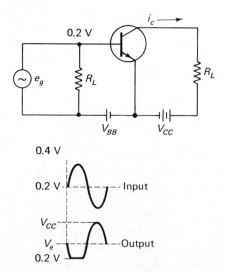

FIGURE 3–15. Common Emitter Driven to Saturation

Here we have a static forward bias of 0.2 V positive between base and emitter. The output is reverse-biased by a positive from battery V_{CC} between collector and emitter. The incoming signal e_g is 0.4 V peak-to-peak amplitude, and it rides on our static 0.2-V bias. As e_g swings positive from 0.2 V toward 0.4 V, our collector current increases and drives the collector potential toward zero. There comes a point in this change when the base is more positive than the collector. At this point, both input and output are forward-biased. A further increase in signal amplitude causes no further increase in i_c, and the output voltage ceases to drop. We can then define saturation as the leveling off of output current when the input bias is equal to the output potential. Notice that the output waveshape has the negative-going peak clipped off. This clipping action is called saturation limiting. If we wish to avoid saturation limiting, we must bias the transistor so that the incoming signal never causes both junctions to become forward-biased. In this case, we could avoid clipping by increasing our static bias to 0.4 V. The increased bias would move our operating point farther from the point of saturation and give us more room for normal amplification.

Signal-handling Ability We must keep in mind that transistor characteristics *do not* remain uniform between the extreme limits of cutoff and saturation. In the common emitter, our amplification factor (beta) changes as i_c changes. This change causes signal distortion when our operating point is set too close to either extreme. Distortion is not limited to the limiting action just described; any change in waveshape between input and output is distortion. If we are to avoid distortion, we must consider the maximum input signal amplitude and plan a suitable bias. Generally we bias the transistor to place the operating point at the mid-position between cutoff and saturation. When biased in this manner, our transistor has its greatest signal-handling ability.

Temperature Effects Temperature affects both the leakage current and the barrier potential of our transistor. An increase in temperature causes an increase in leakage current. This increase in leakage current has a tendency to shift our operating point in the direction of saturation. We find that leakage current will double for each 8°C increase in temperature for a germanium transistor. If the transistor is silicon, we can expect the leakage current to double for each 5°C rise in temperature. We see that germanium is less sensitive to temperature changes, but silicon has less leakage current to start with. The total leakage current in a germanium transistor is about 100 times more than in a similar transistor made of silicon.

Stability Factor The stability factor is a change in collector current divided by the change in leakage current that produces it. This factor helps us to compare the effectiveness of various biasing methods. The equation is

$$S = \frac{\Delta I_c}{\Delta I_{co}}$$

This stability factor tells us how much the change in leakage current will be multiplied. In a common base circuit, the stability factor is approximately 1. This is the best obtainable. In a common emitter circuit, the stability factor is approximately the same as beta. The higher the stability factor, the less stable the circuit. So if we expect to use a common emitter circuit, we must devise a means of minimizing the effect of leakage current. As we lower the effects of leakage current, our operating point becomes more stable.

Types of Stabilization Circuits We find that signals are coupled into an amplifier in two general ways: series and shunt. The type of input coupling has a bearing on the type of biasing circuit that we shall use. A series arrangement is shown in Fig. 3-16.

The input signal is coupled across a transformer to the transistor base. This places the signal in series with the base bias resistor. The V_{CC} supply causes current from ground through R2 and R1. The

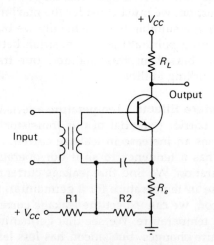

FIGURE 3-16. Series Fed Amplifier

voltage drop across R2 is the dc bias. The capacitor is added to bypass the ac around the biasing network. Our best stability should be realized when R_e is as large as possible and R1 and R2 are as small as possible. But in practical circuits, we have to compromise. In this case, we must trade off some stability to obtain a reasonably small drain on the power supply. So we use moderately large resistors for R1 and R2.

A shunt input bias arrangement is shown in Fig. 3–17. Here we have the signal coupled across a capacitor to the transistor base, and the signal is in shunt with the base bias resistor. The smaller the resistance of R1 and R2, the greater the shunting effect and the larger the signal loss. For minimum signal loss, R1 and R2 should be as large as possible, but for maximum stabilization they should be as small as possible. So now we have a compromise between signal degeneration and stabilization.

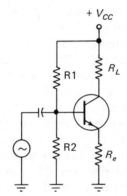

FIGURE 3–17. Shunt Fed Amplifier

In each of the two previous circuits, we used a resistor in the emitter circuit R_e. The total transistor current must pass through R_e; this includes both dc and ac. The polarity on this resistor is such that it follows the potential on the base. For instance, if base potential goes positive, emitter current increases, and emitter goes positive. This following action produces an emitter potential that opposes every change in transistor current. This is a form of degenerative feedback. This emitter degeneration greatly improves stability, but it causes a considerable loss in signal amplification. We can use resistor R_e to oppose slow changes and still obtain maximum signal gain by adding another component to the circuit. This modified circuit is illustrated in Fig. 3–18.

Here we have added a capacitor to bypass the emitter resistor. Now fast changes, such as those produced by the signal, will take

90   Principles of Amplification

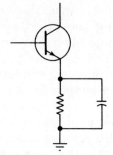

FIGURE 3-18.   Emitter Bypass

a path through the capacitor. These fast changes will cause no change in the emitter potential. As a result, there is no degeneration of the signal. However, the slow changes, such as those produced by temperature effects, will cause a change in the emitter potential. As a result, the slow changes will cause degenerative feedback that still provides a high degree of stabilization.

We may also enhance stabilization by using degenerative feedback from collector to base. One method of collector feedback is illustrated in Fig. 3-19. Our base resistor R_b in this diagram is connected directly between base and collector. This means that the collector potential determines the dc potential on the base. Any increase in collector current reduces the collector voltage and reduces the positive potential on the base. Any decrease in collector current has the opposite reaction. So we see that every change in collector current causes a change in the base potential in a direction that opposes the original change. Unfortunately, this feedback is also detrimental to the signal.

With a slight modification on the circuit in Fig. 3-19, we can obtain considerable stabilization and still retain full signal amplification. We can divide R_b into two resistors and use a capacitor for ac ground between the two resistors. Our circuit is now that in Fig.

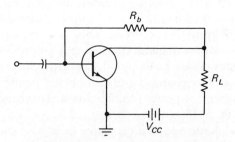

FIGURE 3-19.   Collector Feedback

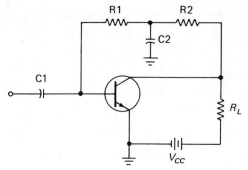

FIGURE 3-20. Improved Collector Feedback

3-20. We still have degenerative feedback for all slow changes, but the fast changes are shunted through the capacitor C2 and do not register on the base potential.

REVIEW EXERCISES

1. What function is performed by an amplifier?
2. Name the three elements of a signal that can be amplified.
3. A circuit is designed to use a PNP transistor as an amplifier. What must be done to the circuit if we use an NPN transistor?
4. The three waveshapes in Fig. 3-21 are outputs from three amplifiers. The inputs in each case are sine waves. Identify the class of amplifier associated with each output.

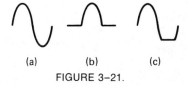

FIGURE 3-21.

5. What conditions dictate the class of amplifier to be used in any given situation?
6. Figure 3-22 contains diagrams of three amplifiers. Identify the circuit configuration of each.
7. Draw a PNP transistor in a common base configuration with proper bias polarities.
8. A given amplifier has a power gain of 30 dB. When the input power is 10 mW, what is the approximate output power?
9. Which of the circuit configurations is associated with the term "alpha"? What is meant by alpha?

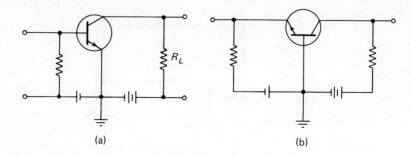

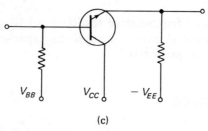

FIGURE 3-22

10. In a certain amplifier the collector current increases 1 mA as the emitter current increases 1.15 mA. Calculate alpha for this amplifier.
11. What elements of the signal is amplified by the common:
    a. Base?
    b. Emitter?
    c. Collector?
12. What is the phase relation of input to output signals in the common:
    a. Base?
    b. Emitter?
    c. Collector?
13. In terms of current gain, what is the cutoff frequency of an amplifier?
14. What causes a transistor's gain to decrease as signal frequency increases?
15. Which of the three amplifier configurations has the highest gain in:
    a. Current?
    b. Voltage?
    c. Power?

16. What information is given on a family of characteristic curves for a common emitter amplifier?
17. A certain common emitter amplifier has dc current of 4 mA in the collector and 100 µA in the base. What is the value of dc beta?
18. Write the equation used to convert

    a. Beta to alpha.
    b. Alpha to beta.
19. A certain transistor has an alpha of 0.95. What is the value of beta for this transistor?
20. A certain transistor in a common emitter circuit has an input resistance of 400 ohms, a beta of 40, and a load resistance of 15,000 ohms. How much voltage gain can be expected from this amplifier?
21. What are the two outstanding characteristics of a common collector?
22. What is the one job that can be performed by the common collector and not by either of the other configurations?
23. Express emitter current in terms of I_c and I_b.
24. A certain common collector circuit has an I_b of 100 µA and an I_c of 4 mA. What is the current gain in this stage?
25. Why is the power gain of a common collector always just a little less than the current gain?
26. What are the extreme limits of a transistor's amplifying action?
27. In a common emitter amplifier, what condition constitutes cutoff?
28. What biasing conditions are present when a transistor is saturated?
29. How should we bias a transistor to avoid limiting due to

    a. Cutoff?
    b. Saturation?
30. Where would we set the operating point to obtain the maximum signal-handling ability from our transistor?
31. What effect does an increase in temperature have on transistor current?
32. What type of feedback is present in the circuit shown in Fig. 3–23?
33. What effect does the feedback in Fig. 3–23 (page 94) have on

    a. Stability?
    b. Signal gain?
34. Add another component to the circuit of Fig. 3–23 which will retain most of the stability without decreasing the signal gain.
35. What type of feedback is illustrated in Fig. 3–24?
36. Redraw the circuit in Fig. 3–24 (page 94) to retain feedback without causing serious loss of signal gain.

94 Principles of Amplification

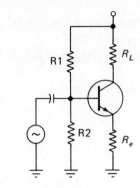

FIGURE 3-23

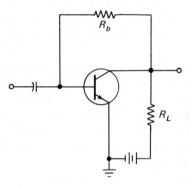

FIGURE 3-24

Chapter Four

SMALL SIGNAL AMPLIFIERS

Small signal amplifier analysis is an area of electronic circuits that is little understood and largely neglected except among design engineers. The approach to these circuits is drastically different from the analysis of more conventional amplifiers, and in the past a full understanding of small signal amplification has not been vital. The serious student of electronics can no longer bypass this area without leaving a serious gap in his training. Technically a small signal is one that has an amplitude that is small in comparison to the dc bias. Our present activities in space, with accompanying improvements in electronic equipment, cause a great many of our signals to fall into the small signal category.

DYNAMIC CHARACTERISTICS

In previous discussions we have analyzed transistor characteristics from a static point of view. These static characteristics tell us the maximum capabilities of a transistor, but not what it will do in a particular circuit. This information can be obtained only by converting the static characteristics to dynamic characteristics. Dynamic characteristics are calculated by assuming or measuring specific circuit configurations and operating conditions.

The conventional method of developing dynamic characteristics is to draw a load line on a family of static characteristic curves. The load line works very well for large signal amplifiers, and we shall use them in Chapter 5. However, small signals *do not* produce enough variation in the current and voltage to make the load line analysis practical. But we have another method of determining the dynamic characteristics that can be applied to small signal amplifiers.

96 Small Signal Amplifiers

Equivalent Circuits We can calculate the dynamic characteristics of a circuit by constructing an equivalent circuit and using variable quantities of current and voltage. Working with symbols, we can calculate the resistance, conductance, and gain of current, voltage, and power of an amplifier without knowing the exact values. We refer to the quantities voltage, current, conductance, and resistance in our equivalent circuits as "parameters." A parameter is simply a quantity whose value varies with the circumstances of its application. We use symbols for the parameters and do most of our calculations apart from actual values.

We shall use the common emitter amplifier to develop our equivalent circuits and parameters. This is a convenient configuration, the most used circuit, and we can convert the parameters to the other configurations when necessary. Figure 4-1 is a common emitter that illustrates our variable quantities of voltage and current.

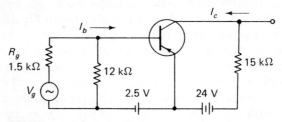

FIGURE 4-1. Common Emitter with Variable Quantities

Even when we have a given R_g and a given R_L, we still have four variables of current and voltage. These are

  i_b input current
  V_{be} input voltage
  i_c output current
  V_{bc} output voltage

Handling four variables simultaneously presents impractical complications in mathematics. So we simplify the problem by opening or shorting two of the variables. (The opens and shorts referred to in this chapter are open and short to ac only. The dc circuit will still be complete.) The two that we open or short become independent variables, and the remaining two are dependent variables. We develop open-circuit parameters, short-circuit parameters, or a mixture of open and short (hybrid) parameters according to our selection of dependent and independent variables.

Since a transistor in the common emitter configuration amplifies both voltage and current, we may consider it to be two separate generators. In other words, it effectively generates both voltage and current. When we construct our equivalent circuit, we replace the transistor with a voltage generator and a current generator. Figure 4-2 is an equivalent circuit for the diagram in Fig. 4-1.

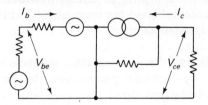

FIGURE 4-2. Equivalent Circuit

We represent the voltage generator with a single circle in series with a resistor and the current generator with a double circle in parallel with a resistor. These two generators incorporate all the dynamic characteristics of our transistor. Proper analysis of these equivalent elements will reveal all that we need to know about how our amplifier will react to the conditions of application.

Black Box If we remove the input and output circuit from the diagram in Fig. 4-1, we have a transistor with four leads. If we enclose the transistor in a box, as shown in Fig. 4-3, we have no way of knowing if the box contains an actual transistor or the generators shown in our equivalent circuit. We shall use this black box concept to develop our parameters.

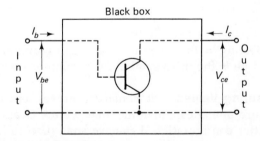

FIGURE 4-3. Variable Quantities

OPEN-CIRCUIT PARAMETERS

When we choose the two currents as our independent variables, we can determine the resistance parameters. We must construct test circuits and take certain measurements for use in our calculations. Because of the configuration of the test circuits, we refer to the resistance parameters as open-circuit parameters; the measurements are made under specified open-circuit conditions. Remember that we are dealing with opens only so far as ac is concerned. Obviously, a circuit must have a complete dc circuit in order to function. The ac open can be accomplished by connecting a choke coil in series with the lead designated as open.

Designations We use specific symbols to designate all parameters. With open-circuit parameters, the prime symbol is r, which designates resistance. Some texts deal with impedance instead of resistance. If impedance is used as the open-circuit parameters, then the primary symbol z is used instead of r. We are interested in four specific resistances, and we combine r with another symbol to specify which resistance we are concerned with. Here are the four open-circuit parameters with the identifying symbols:

r_i = input resistance with output open.
r_r = reverse transfer resistance with input open.
r_f = forward transfer resistance with output open.
r_o = output resistance with input open.

A third symbol is added to the first two to designate the circuit configuration of the amplifier. In our case, we are using the common emitter circuit, so our third symbol is e. We simply add the e to each of our previous symbols as follows:

$$r_{ie}$$
$$r_{re}$$
$$r_{fe}$$
$$r_{oe}$$

If we were dealing with a common base configuration, our third symbol would be b; the third symbol for a common collector is c.

Measuring Values We construct our test circuits as illustrated in Figure 4-4. Keep in mind that we are dealing with the common emitter configuration. Later we shall discuss how parameters can be converted among the three configurations. We shall also discuss the conversion among open, short, and hybrid parameters for a given configuration.

With test circuit a, we are determining the input resistance with the output open r_{ie}. This leaves us i_b and V_{be} to work with, and we measure the change in each of these quantities. This change ratio is r_{ie}, and it is expressed as

$$r_{ie} = \frac{\Delta V_{be}}{\Delta I_b}$$

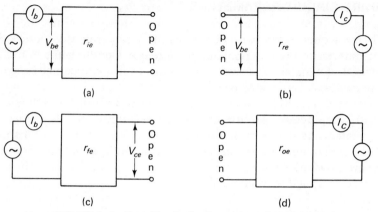

FIGURE 4-4. Test Circuits for Open-circuit Parameters

With test circuit b, we are concerned with the reverse transfer resistance with an open input r_{re}. Now we work with the ratio of change between V_{be} and I_c, which we express as

$$r_{re} = \frac{\Delta V_{be}}{\Delta I_c}$$

Test circuit c is used to determine the forward transfer resistance with an open output r_{fe}. This r_{fe} is the change ratio between V_{ce} and I_b. As an equation, it is

$$r_{fe} = \frac{\Delta V_{ce}}{\Delta I_b}$$

We use test circuit d to determine the output resistance when the input is open r_{oe}. We now work with V_{ce} and I_c and calculate the change ratio between these two. This ratio is expressed as

$$r_{oe} = \frac{\Delta V_{ce}}{\Delta I_c}$$

Look back at each of the open circuit equations, and notice that each parameter is obtained by a simple application of Ohm's law,

$$R = \frac{E}{I}$$

In each case, we measure the value of E and I and use these values to calculate R.

SHORT-CIRCUIT PARAMETERS

When we choose the two voltages as the independent variables, we can determine the conductance parameters. Conductance is the reciprocal of resistance, so we should expect the conductance formulas to be reciprocals of the resistance formulas. If we had used impedance as the open-circuit parameters, we would use admittance as the short-circuit parameters. Admittance is the reciprocal of impedance, as conductance is the reciprocal of resistance. The test circuits for short-circuit parameters are constructed and shorted in a specified manner. We are now referring to ac shorts, and we accomplish these by connecting a capacitor in parallel.

Designators With short-circuit parameters we use the symbol g to designate conductance. If admittance were used instead of conductance, the symbol would be y. The other symbols that we combine with g have the same meaning as before. Our four short-circuit parameters are

g_i = input conductance with output shorted.
g_r = reverse transfer conductance with input shorted.
g_f = forward transfer conductance with output shorted.
g_o = output conductance with input shorted.

Again we add a third symbol of b, e, or c to designate the circuit configuration. With our common emitter, the symbols are

$$g_{ie}$$
$$g_{re}$$
$$g_{fe}$$
$$g_{oe}$$

Measuring Values We construct the test circuits for short-circuit parameters as illustrated in Fig. 4–5. There is a separate test circuit arrangement for each of our parameters. Notice the position of the short in each case.

Test circuit a provides the values for determining the input conductance when the output is shorted g_{ie}. We measure the change in I_b and V_{be} and calculate as follows:

$$g_{ie} = \frac{\Delta I_b}{\Delta V_{be}}$$

Test circuit b provides the information that we need to calculate the reverse transfer conductance with a shorted input g_{re}. We

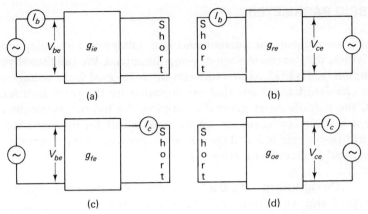

FIGURE 4-5. Test Circuits for Short-circuit Parameters

now work with V_{ce} and I_b, and since this is reverse conductance, the I_b is a negative value. The equation is expressed as

$$g_{re} = \frac{\Delta - I_b}{\Delta V_{ce}}$$

We use test circuit c to determine the forward transfer conductance with the output shorted g_{fe}. We measure the change in I_c and V_{be} and use the changes in this equation:

$$g_{fe} = \frac{\Delta I_c}{\Delta V_{be}}$$

Test circuit d provides the data we need for calculating the output conductance when the input is shorted g_{oe}. This quantity is the change ratio between I_c and V_{ce}. The equation is

$$g_{oe} = \frac{\Delta I_c}{\Delta V_{ce}}$$

In each of the short-circuit equations, we divided current by voltage, because conductance is the reciprocal of resistance. Each equation is a simple application of the conductance formula

$$G = \frac{I}{E}$$

In each case, we measure the values of I and E and use simple mathematics to calculate G.

HYBRID PARAMETERS

When we choose one current and one voltage as our independent variables, the parameters we develop are mixed. We call them hybrid parameters. Hybrid parameters can be developed from voltage-current characteristic charts that are supplied by the manufacturer. In fact, the manufacturer generally supplies the hybrid parameters for one amplifier configuration on the data sheet for the transistor. A hybrid parameter is a voltage-current ratio with certain other quantities held at a constant value.

Designators We use h as a symbol to designate hybrid parameters, and we have four of them. We still use $i, r, f,$ and o as secondary designators to show input, reverse, forward, and output, respectively. The four hybrid parameters are

h_i = hybrid input resistance.
h_r = hybrid reverse voltage gain.
h_f = hybrid forward current gain.
h_o = hybrid output conductance.

The symbols $b, e,$ or c are added as before to designate the type of circuit configuration. For our common emitter, the complete hybrid symbols are

$$h_{ie}$$
$$h_{re}$$
$$h_{fe}$$
$$h_{oe}$$

Input Resistance h_{ie} Input resistance is the ratio of change between V_{be} and I_b when we hold V_{ce} to a constant value. These are incremental (small) changes, and the equation becomes

$$h_{ie} = \frac{\Delta V_{be}}{\Delta I_b} \quad (V_{ce} \text{ constant})$$

The changes may be measured or taken from a graph of characteristics similar to that illustrated in Fig. 4–6.

First we establish bias by specifying the values of base current I_b and the voltage between collector and emitter V_{ce}. These values should be selected to give us an operating point centered on the most linear portion of the voltage-current curve. Let us specify $I_b = 150 \mu A$ and $V_{ce} = 7.5$ V. This places our operating point at point X on the graph. Now we hold V_{ce} constant at 7.5 V and make a small change in either V_{be} or I_b. If we change V_{be}, it causes a change in I_b and vice versa. The ratio of these changes is h_{ie}.

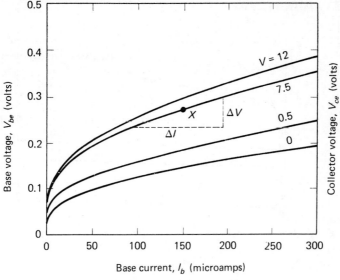

FIGURE 4-6. Calculating h_{ie}

Let us change I_b by 50 μA to either side of point X. This change is plotted on the graph, and ΔI_b is 100 μA. This change of I_b causes a change in V_{be} from 0.22 V to 0.28 V; ΔV_{be} is 0.06 V. Substituting these values into the equation for h_{ie}, we can solve for input resistance:

$$h_{ie} = \frac{\Delta V_{be}}{\Delta I_b}$$

$$= \frac{0.06 \text{ V}}{100 \text{ μA}}$$

$$= \frac{0.6 \times 10^6}{100}$$

$$= 600 \text{ Ω}$$

Reverse Voltage Gain h_{re} Forward voltage gain, in all cases, is E_{out}/E_{in}. The reverse voltage gain is a reciprocal of forward voltage gain, so it is E_{in}/E_{out}. Our reverse voltage gain is calculated by dividing a small change in base voltage by the resulting small change in collector voltage. Using a constant value of base current, we determine these changes from a graph. The equation is

$$h_{re} = \frac{\Delta V_{be}}{\Delta V_{ce}} \quad (I_b \text{ constant})$$

The graph for plotting the changes is similar to that in Fig. 4-7.

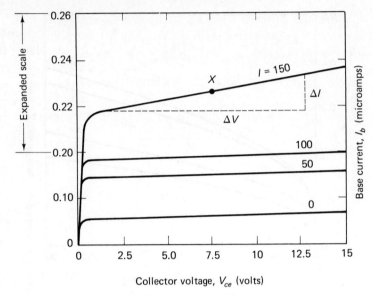

FIGURE 4-7. Calculating h_{re}

Using the same starting point as before, we have the operating point at $I_b = 150\ \mu A$ and $V_{ce} = 7.5$ V. Now we change V_{be} by a small amount and observe the change it causes in V_{ce}. Let us move V_{be} down to 0.22 V and back up to 0.235 V. This change of 0.015 V in V_{be} causes V_{ce} to change from 2.5 V to 12.5 V; a total of 10 V. We can substitute these values into the h_{re} equation and solve for the reverse voltage gain, as follows:

$$h_{re} = \frac{\Delta V_{be}}{\Delta V_{ce}}$$

$$= \frac{0.015\ V}{10\ V}$$

$$= 0.0015 = 15 \times 10^{-4}$$

The reverse voltage gain is an undesirable characteristic of the transistor. It represents internal feedback of a degenerative nature, which tends to reduce the forward voltage gain. In fact, we may consider reverse voltage gain and forward voltage gain to be reciprocals of each other. We can then compute the forward voltage gain as follows:

$$A_v\ (\text{forward}) = \frac{1}{h_{re}}$$

Forward Current Gain h_{fe} Forward current gain is an incremental change ratio between I_c and I_b with V_{ce} constant. The equation is

$$h_{fe} = \frac{\Delta I_c}{\Delta I_b} \quad (V_{ce} \text{ constant})$$

The graph in Fig. 4–8 provides the information necessary for calculating h_{fe}.

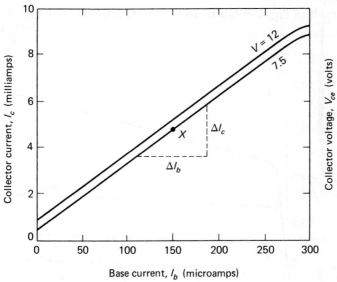

FIGURE 4–8. Calculating h_{fe}

If we use the same starting conditions as before ($I_b = 150$ μA and $V_{ce} = 7.5$ V), our operating point is indicated by point X on the graph. If we change I_b by 50 μA, it moves down to 125 μA and back up to 175 μA. This causes I_c to change by 1 mA. It moves down to 4.5 mA and back up to 5.5 mA. We can use the values to solve for the forward current gain as follows:

$$h_{fe} = \frac{\Delta I_c}{\Delta I_b}$$

$$= \frac{1 \text{ mA}}{50 \text{ }\mu\text{A}}$$

$$= \frac{1 \times 10^{-3}}{50 \times 10^{-6}}$$

$$= 20$$

106 Small Signal Amplifiers

Output Conductance h_{oe} Since output conductance is the reciprocal of output resistance, this is another calculation that performs double duty for us. Output conductance is an incremental change ratio between I_c and V_{ce} with I_b held at a constant value. The equation is

$$h_{oe} = \frac{\Delta I_c}{\Delta V_{ce}} \quad (I_b \text{ constant})$$

The graph in Fig. 4–9 can be used to obtain the incremental changes.

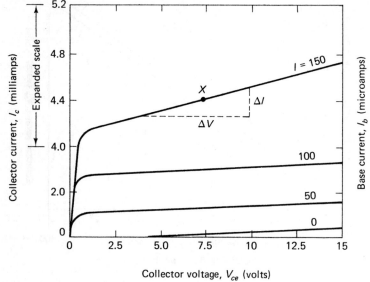

Collector voltage, V_{ce} (volts)

FIGURE 4–9. Calculating h_{oe}

Keeping our familiar operating point ($I_b = 150\ \mu A$ and $V_{ce} = 7.5$ V), let us change I_c and observe the resulting change in V_{ce}. If we change I_c by 0.2 mA, it moves down to 4.3 mA and back up to 4.5 mA. This changes V_{ce} by 6 V. It moves down to 4.5 V and up to 10.5 V. Using these values, we calculate as follows:

$$h_{oe} = \frac{\Delta I_c}{\Delta V_{ce}}$$

$$= \frac{0.2 \text{ mA}}{6 \text{ V}}$$

$$= 33 \times 10^{-6}$$

$$= 33\ \mu\mho$$

The term $\mu\mho$ means "micromhos," which is the usual expression for output conductance. Since this is the reciprocal of output resistance, we can express output resistance as

$$r_{out} = \frac{1}{h_{oe}}$$

Other Configurations The calculations for hybrid parameters are much the same in all three circuit configurations. Since we have developed the procedures in some detail for the common emitter, it would be unduly repetitious to do the same for the common base and the common collector. However, here are the equations for these other configurations. They may be useful.

COMMON BASE

$h_{ib} = \dfrac{\Delta V_{be}}{\Delta I_e}$ (V_{cb} constant)

$h_{rb} = \dfrac{\Delta V_{eb}}{\Delta V_{cb}}$ (I_e constant)

$h_{fb} = \dfrac{\Delta I_c}{\Delta I_e}$ (V_{cb} constant)

$h_{ob} = \dfrac{\Delta I_c}{\Delta V_{cb}}$ (I_e constant)

COMMON COLLECTOR

$h_{ic} = \dfrac{\Delta V_{bc}}{\Delta I_b}$ (V_{ec} constant)

$h_{rc} = \dfrac{\Delta V_{bc}}{\Delta V_{ec}}$ (I_b constant)

$h_{fc} = \dfrac{\Delta I_e}{\Delta I_b}$ (V_{ec} constant)

$h_{oc} = \dfrac{\Delta I_e}{\Delta V_{ec}}$ (I_b constant)

The hybrid parameter values vary over a wide range as we consider different transistors. For this reason, there are no typical values for these parameters. With a given transistor, there is a wide difference in parameters when we place the transistor in different circuit configurations. Here is a comparison of hybrid parameter values of a given transistor in all three configurations.

COMMON BASE	COMMON EMITTER	COMMON COLLECTOR
$h_{ib} = 39\ \Omega$	$h_{ie} = 1900\ \Omega$	$h_{ic} = 1950\ \Omega$
$h_{rb} = 380 \times 10^{-6}$	$h_{re} = 575 \times 10^{-6}$	$h_{rc} = 1$
$h_{fb} = -0.98$	$h_{fe} = 49$	$h_{fc} = -50$
$h_{ob} = 0.49\ \mu\mho$	$h_{oe} = 24\ \mu\mho$	$h_{oc} = 24.5\ \mu\mho$

Converting Parameters among Configurations In most cases, we can obtain one set of hybrid parameters from the manufacturer's data sheet. If we want to use the transistor in a different configuration, we need to convert the given parameters to fit the cir-

108 Small Signal Amplifiers

cuit of our choice. The following equations can be used for this conversion. The equality signs in these equations should be interpreted loosely. Mathematically the quantities on either side of the equality sign *are not* always exactly the same, but they are close enough for most practical applications.

FROM CE TO CB

$$h_{ib} = \frac{h_{ie}}{1 + h_{fe}}$$

$$h_{rb} = \frac{(h_{ie})(h_{oe})}{1 + h_{fe}} - h_{re}$$

$$h_{fb} = \frac{-h_{fe}}{1 + h_{fe}}$$

$$h_{ob} = \frac{h_{oe}}{1 + h_{fe}}$$

FROM CE TO CC

$$h_{ic} = h_{ie}$$

$$h_{rc} = 1 - h_{re}$$

$$h_{fc} = (1 + h_{fe})$$

$$h_{oc} = h_{oe}$$

FROM CB TO CE

$$h_{ie} = \frac{h_{ib}}{1 + h_{fb}}$$

$$h_{re} = \frac{(h_{ib})(h_{ob})}{1 + h_{fb}} - h_{rb}$$

$$h_{fe} = \frac{-h_{fb}}{1 + h_{fb}}$$

$$h_{oe} = \frac{h_{ob}}{1 + h_{fb}}$$

FROM CB TO CC

$$h_{ic} = \frac{h_{ib}}{1 + h_{fb}}$$

$$h_{rc} = \frac{(h_{ib})(h_{ob})}{1 + h_{fb}} - h_{rb}$$

$$h_{fc} = \frac{-1}{1 + h_{fb}}$$

$$h_{oc} = \frac{h_{ob}}{1 + h_{fb}}$$

Suppose that we have a transistor with an h_{ie} of 1950 Ω and an h_{fe} of 49, but we want to use it in a common base amplifier. What is the value of h_{ib}?

$$h_{ib} = \frac{h_{ie}}{1 + h_{fe}}$$

$$= \frac{1950 \, \Omega}{1 + 49}$$

$$= \frac{1950}{50}$$

$$= 39$$

Let us try another. Assume that we have a transistor with an h_{ob} of 0.49 $\mu\mho$ and an h_{fb} of -0.98, but we want to use it as a common collector. What is the value of h_{oc}?

$$h_{oc} = \frac{h_{ob}}{1 + h_{fb}}$$

$$= \frac{0.49\ \mu\mho}{1 + (-98)}$$

$$= \frac{0.49 \times 10^{-6}}{0.02}$$

$$= 24.4\ \mu\mho$$

PARAMETER INTERRELATIONSHIP

There is a relationship among open-circuit, short-circuit, and hybrid parameters. You already know that the open- and short-circuit parameters are exact reciprocals of each other. The open-circuit parameters are E/I for resistance, and the short-circuit parameters are I/E for conductance. A little more mathematics is involved, but we can also convert from hybrid to short-circuit parameters and vice versa, or from hybrid to open-circuit parameters or vice versa.

We shall use a common emitter circuit and illustrate the procedure for converting. We shall convert from hybrid to open circuit and to short circuit. Then we shall convert from open circuit back to hybrid. The formulas and procedures are the same for all configurations except for the configuration designator. Therefore, it is not necessary to go into the common base or the common collector circuits.

From Hybrid to Open Circuit Remember that our open-circuit parameters are resistance values. Here are the equations for converting from hybrid to open circuit.

$$r_{ie} = h_{ie} - \frac{(h_{re})(h_{fe})}{h_{oe}} \qquad r_{fe} = \frac{-h_{fe}}{h_{oe}}$$

$$r_{re} = \frac{h_{re}}{h_{oe}} \qquad r_{oe} = \frac{1}{h_{oe}}$$

Since all of our hybrid parameters would be known quantities, the procedure is simple substitution and a little algebra. For example, suppose that we have the following hybrid parameters:

$$h_{ie} = 39\ \Omega$$
$$h_{re} = 380 \times 10^{-6}$$
$$h_{fe} = -98$$
$$h_{oe} = 0.49 \times 10^{-6}$$

110 Small Signal Amplifiers

To solve for r_{ie}, we proceed as follows:

$$r_{ie} = h_{ie} - \frac{(h_{re})(h_{fe})}{h_{oe}}$$

$$= \frac{39 - (380 \times 10^{-6})(-98)}{0.49 \times 10^{-6}}$$

$$= 39 - (-760)$$

$$= 799 \, \Omega$$

From Hybrid to Short Circuit Our short-circuit parameters are conductance values. The equations for converting hybrid parameters to short-circuit parameters are as follows:

$$g_{ie} = \frac{1}{h_{ie}} \qquad g_{fe} = \frac{h_{fe}}{h_{ie}}$$

$$g_{re} = \frac{-h_{re}}{h_{ie}} \qquad g_{oe} = h_{oe} - \frac{(h_{re})(h_{fe})}{h_{ie}}$$

Let us use the same assumed values for our hybrid parameters and solve for g_{fe}.

$$g_{fe} = \frac{h_{fe}}{h_{ie}}$$

$$= \frac{-98}{39}$$

$$= -2.5 \text{ mhos}$$

From Open Circuit to Hybrid Since open- and short-circuit parameters are reciprocals of each other, we shall convert only from open circuit to hybrid parameters. Our equations are

$$h_{ie} = r_{ie} - \frac{(r_{re})(r_{fe})}{r_{oe}} \qquad h_{fe} = -\frac{r_{fe}}{r_{oe}}$$

$$h_{re} = \frac{r_{re}}{r_{oe}} \qquad h_{oe} = \frac{1}{r_{oe}}$$

Assume that we have the following open-circuit parameters:

$$r_{re} = 776 \, \Omega$$

$$r_{oe} = 2 \text{ M}\Omega$$

Using these figures to solve for h_{re}, we have

$$h_{re} = \frac{r_{re}}{r_{oe}}$$

$$= \frac{776 \, \Omega}{2 \, \text{M}\Omega}$$

$$= \frac{776 \times 10^{-6}}{2}$$

$$= 388 \times 10^{-6}$$

Small Signal Characteristics The parameters that we have been discussing are small signal characteristics. Some data sheets use the same symbols that you have seen here. However, there are exceptions to this rule. Most frequently encountered of these exceptions are

$$h_{11}$$
$$h_{12}$$
$$h_{21}$$
$$h_{22}$$

These are hybrid parameters, and we relate them as follows:

	CB	CE	CC
h_{11} = input resistance	= h_{ib}	h_{ie}	h_{ic}
h_{12} = reverse voltage gain	= h_{rb}	h_{re}	h_{rc}
h_{21} = forward current gain	= h_{fb}	h_{fe}	h_{fc}
h_{22} = output conductance	= h_{ob}	h_{oe}	h_{oc}

When alpha and beta are not specified in plain terms:

$$h_{11} = h_{ib} = \text{alpha for CB}$$
$$= h_{ie} = \text{beta for CE}$$

Other variations that we may expect on the data sheets are

$$\alpha_f$$
$$\mu_r$$

These are also hybrid parameters and:

$$\alpha_f = \text{forward current gain}$$
$$\mu_r = \text{reverse voltage gain}$$

APPLYING PARAMETERS

Now that we have developed the various parameters, let us see what practical applications we have for them. We should be able to predict the behavior of a transistor under various operating conditions.

Equivalent Circuits We can develop equivalent circuits to fit any transistor and any circuit configuration. In terms of parameters, we can then establish values for all parts of the circuit. For example, consider the circuit in Fig. 4-10.

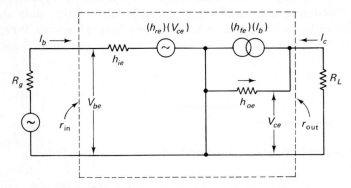

FIGURE 4-10. Equivalent Circuit with Parameters

We can now use the hybrid parameter symbols from the equivalent circuit and construct our own equations for current gain, voltage gain, and power gain. The equations will be valid for any transistor and any values of R_g, R_L, and bias voltages. When we determine the actual values, we can substitute these into the equations, and the rest is simple algebra. To develop these equations, we simply take our original formulas and substitute parameters from the equivalent circuit.

Current Gain

$$A_i = \frac{I_c}{I_b}$$

becomes

$$A_i = \frac{h_{fe}}{(h_{oe})(R_L) + 1}$$

Voltage Gain

$$A_r = \frac{V_{ce}}{V_{be}}$$

becomes

$$A_r = \frac{(-h_{fe})(R_L)}{[(h_{ie})(h_{oe}) - (h_{fe})(h_{re})][R_L + (h_{ie})]}$$

Power Gain

$$G = \frac{I_c V_{ce}}{I_b V_{cb}}$$

becomes

$$G = \frac{(h_{fe})^2 R_L}{(h_{oe})(R_L + 1)[(h_{ie})(h_{oe}) - (h_{fe})(h_{re})][R_L + h_{ie}]}$$

REVIEW EXERCISES

1. What type of general information concerning a transistor's performance can be obtained from the
   a. Static characteristics?
   b. Dynamic characteristics?
2. Name two methods of converting static to dynamic characteristics.
3. Which of the methods in item 2 is applicable to small signal amplifiers? Why?
4. Define the term "parameter."
5. What are the four variables that we use to develop parameters?
6. Why do we use voltage and current generator symbols to represent a transistor in an equivalent circuit?
7. What symbol is used in equivalent circuits to represent a
   a. Voltage generator?
   b. Current generator?
8. Why are the resistance parameters called open-circuit parameters?
9. List the symbols for open-circuit parameters for a common base amplifier.
10. Write the meaning of the first two letters in each of the symbols referred to in item 9.

114 Small Signal Amplifiers

11. Which letter of all parameter symbols designates the circuit configuration? What is this letter for the
    a. Common base?
    b. Common emitter?
    c. Common collector?

12. In a common emitter configuration, which parameter is indicated by each of the following ratios?

    a. $\dfrac{\Delta V_{be}}{\Delta I_b}$ b. $\dfrac{\Delta V_{be}}{\Delta I_c}$ c. $\dfrac{\Delta V_{ce}}{\Delta I_b}$ d. $\dfrac{\Delta V_{ce}}{\Delta I_c}$

13. What characteristics are we interested in when we deal with short-circuit parameters?

14. Write the symbols for the short-circuit parameters for a common emitter amplifier.

15. Write the meaning of the first two letters in each of the symbols referred to in item 14.

16. Assuming a common emitter configuration, match a short-circuit parameter symbol to each of these ratios:

    a. $\dfrac{\Delta I_b}{\Delta V_{be}}$ b. $\dfrac{\Delta -I_b}{\Delta V_{ce}}$ c. $\dfrac{\Delta I_c}{\Delta V_{be}}$ d. $\dfrac{\Delta I_c}{\Delta V_{ce}}$

17. Where do we get the term "hybrid parameter"?

18. Write the symbols for the hybrid parameters of a common emitter configuration.

19. What is the meaning of the first two letters in each of the symbols referred to in item 18?

20. Assuming a common emitter configuration, match a hybrid parameter symbol to each of these ratios:

    a. $\dfrac{\Delta V_{be}}{\Delta I_b}$ (V_{ce} constant) b. $\dfrac{\Delta V_{be}}{\Delta V_{ce}}$ (I_b constant)

    c. $\dfrac{\Delta I_c}{\Delta I_b}$ (V_{ce} constant) d. $\dfrac{\Delta I_c}{\Delta V_{ce}}$ (I_b constant)

21. What is the meaning of each of the following small signal characteristics?

    h_{11}
    h_{12}
    h_{21}
    h_{22}
    α_f
    μ_r

Chapter Five

Large Signal Amplifiers

"Large signal amplifiers" is another nebulous grouping of amplifiers. It is difficult to say exactly where small signal amplifiers stop and large signal amplifiers begin. The best we can do, by way of clarification, is to say that the large signal amplifier has relatively large signals when we compare the signal amplitude to the circuit bias. In other words, when the signal causes enough change to be accurately plotted on a graph, we may elect to place our amplifier in the large signal category for the purpose of analysis.

A large signal amplifier lends itself to a graphical analysis for converting the static characteristics to dynamic characteristics. The graphical analysis includes load lines, dynamic transfer characteristics, and constant power dissipation curves. Many people prefer this method of analysis to the parameters previously discussed. The method has the advantages of being easy to operate, involving less mathematics, and being generally more accurate.

LOAD LINES

A load line is a line superimposed on a family of static characteristic curves in such a way that it shows the dynamic characteristics of a particular circuit. A given transistor reacts differently when we change its loading conditions. For this reason, a load line is accurate for only one specified set of conditions. If we alter the applied voltage or the load resistance, we have a different circuit, and a new load line is required. We can construct load lines for any type of transistor and for any circuit configuration. All we need is a family of static characteristic to match the circuit configuration that we plan to use.

116 Large Signal Amplifiers

Constructing a Load Line   Figure 5–1 is a bipolar transistor in a common emitter amplifier configuration. This illustration provides all the information that we need for constructing a load line.

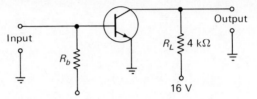

FIGURE 5–1. Common Emitter Amplifier

As we examine the circuit, we find very little information. In fact, all that we have is the output bias and the size of the load resistor. But that, combined with what we already know, is all we need. We would know the type of transistor and would have in our possession a graph of its static characteristics. Let us assume that the graph in Fig. 5–2 matches the transistor in Fig. 5–1.

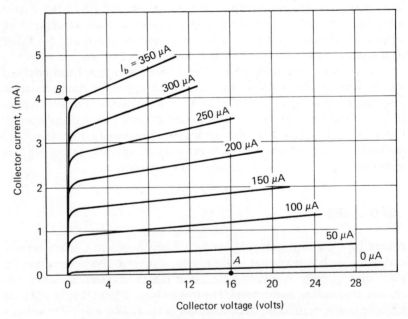

FIGURE 5–2. Common Emitter Characteristics

The first step in constructing the load line is to determine the points of maximum collector current and maximum collector voltage. When we have zero collector current, there is no voltage drop across the load resistor, and the full value of the output bias voltage appears at the collector. The output bias then is the maximum collector voltage. In our case, we have 16 volts for output bias. Notice on the characteristics graph of Fig. 5–2, we have already plotted a dot (point A on the graph) at the point that shows zero collector current and 16-V collector voltage. The point of maximum collector current is determined by dividing the output bias voltage by the value of the load resistor. In this case, we have 16 V/4 kΩ, and this gives us an I_c of 4 mA. At the point of maximum current, the total output bias voltage is dropped across the load resistor, and this gives us zero volts for collector voltage. Point B on the graph of Fig. 5–2 is the maximum current point for this circuit. So point A on our graph represents maximum V_c and minimum (zero) I_c, and point B represents maximum I_c and minimum (zero) V_c. The load line is completed by drawing a straight line between points A and B, as shown in Fig. 5–3.

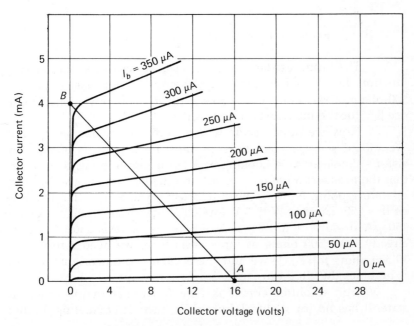

FIGURE 5–3. Completing the Load Line

If you will examine the load line carefully, you will notice that it is a locus of the values of I_c for all possible values of V_c. Of course, it applies only to the circuit shown in Fig. 5–1. If we want to consider a different circuit, we shall draw a different load line. But in this specific circuit, all possible values of I_c and V_c intersect at the load line.

118   Large Signal Amplifiers

Establishing the Operating Point   The operating point is a quiescent state. It is located on the graph at the point where the dc base current intersects the load line. We can establish the operating point at any desired place along the load line by selecting the proper level of input bias. Notice in Fig. 5-1 that we have not designated the values for either input bias or R_b. We are free to select these values so that we can have a specific quantity of base current when there is no signal input.

This static base current is determined by three factors: the resistance of R_b, the input resistance of the transistor, and the quantity of input bias. This input bias circuit is illustrated in Fig. 5-4.

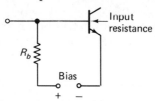

FIGURE 5-4. Factors Affecting Static Base Current

The input resistance is the resistance across the emitter-base junction. Let us assume that it is 500 ohms. In our bias circuit, this 500 ohms of resistance is in series with R_b. For convenience, let us use 3 V for the input bias supply voltage.

Now we must decide where we want our operating point. We should select a point near the most linear portion of the transistor's characteristics. Looking back at the graph in Fig. 5-3, we see that the intersection of $I_b = 150$ μA and the load line is a satisfactory operating point. Now we need to select a value of R_b that will give us this 150 μA of I_b. With 3 V applied to the input circuit, we obtain 150 μA of current if we use 20,000 ohms of resistance. Since we already have 500 ohms of input resistance, we need a resistor of 19,500 ohms for R_b. If we add these values to our original circuit, it becomes the circuit in Fig. 5-5.

Our amplifier circuit is now functional in the quiescent state. It has no input signal, but we do have dc circulating through

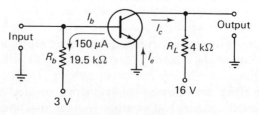

FIGURE 5-5.   Updated Circuit

the transistor. Now let us plot the operating point on our graph. It is located at the point of intersection of the 150-μA curve and the load line. It appears as point C on the graph in Fig. 5–6.

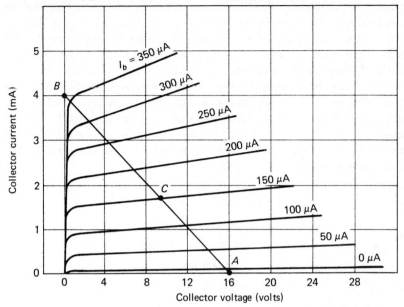

FIGURE 5–6. Operating Point

If we project a vertical line downward from the operating point, we find that the collector has a voltage of 9.2 V. Projecting a horizontal line to the left from the operating point reveals that we have a collector current of 1.7 mA. We can check I_c against V_c by multiplying I_c times R_L.

$$VR_L = I_c R_L$$
$$= 1.7 \text{ mA} \times 4 \text{ k}\Omega$$
$$= 1.7 \times 10^{-3} \times 4 \times 10^3$$
$$= 1.7 \times 4$$
$$= 6.8 \text{ V}$$

The voltage at the collector must be the total output bias supply voltage less the voltage across R_L. Therefore, V_c is 16 V−6.8 V = 9.2 V.

Predicting the Output Any signal applied to the input will cause voltage and current values to vary to either side of the operating point. If we know the amplitude of an input signal, we can predict the exact output in terms of voltage, current, and waveshape. Suppose that our circuit has an input sine wave signal that is 100 μA in (peak-to-peak) amplitude. Let us plot the waveshapes on our graph. When we finish this plot, our graph will look like that in Fig. 5-7.

Our input signal rides the 150 μA of dc base current and causes the base current to swing from 150 μA to 200 μA, back to 150 μA, down to 100 μA, and back up to 150 μA. What amplitude of signal voltage is required for this change in base current? Remember that we have an input resistance of 500 ohms. Now take another look at the circuit in Fig. 5-4. Notice that R_b and input resistance are in series for the dc bias, but they are in parallel for the ac signal. This means that our total signal voltage appears across the junction resistance. The signal voltage then is

$$\begin{aligned} V &= IR \\ &= 100 \; \mu A \times 500 \; \Omega \\ &= 100 \times 10^{-6} \times 500 \\ &= 50{,}000 \times 10^{-6} \\ &= 50 \text{ mV} \end{aligned}$$

We now mark points D and E on our graph. These are the points where maximum i_b and minimum i_b intersect the load line. Points C, D, and E are projected to the left for collector current, and downward for the collector voltage. We find that the collector current follows the input current, and it swings from 1.7 mA to 2.3 mA, to 1.2 mA, and back to 1.7 mA. Our amplifier increases the 100-μA input current to 1.1 mA of output current. This is a current amplification of

$$\begin{aligned} A_i &= \frac{i_{\text{out}}}{i_{\text{in}}} \\ &= \frac{1.1 \text{ mA}}{100 \; \mu A} \\ &= \frac{1.1 \times 10^{-3}}{100 \times 10^{-6}} \\ &= \frac{1.1}{0.1} \\ &= 11 \end{aligned}$$

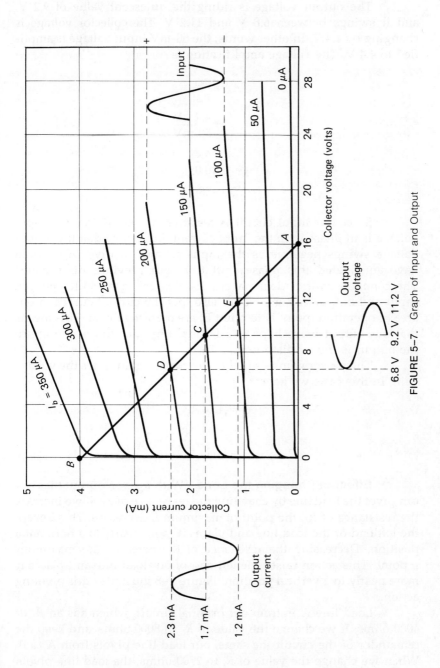

FIGURE 5-7. Graph of Input and Output

The output voltage is riding the quiescent value of 9.2 V, and it swings between 6.8 V and 11.2 V. The collector voltage is changing by 4.4 V. In other words, the 50-mV input voltage is amplified to 4.4 V. The voltage amplification is

$$A_v = \frac{v_{\text{out}}}{v_{\text{in}}}$$

$$= \frac{4.4 \text{ V}}{50 \text{ mV}}$$

$$= \frac{4.4}{0.05}$$

$$= 88$$

Since our input circuit is resistive, we know that the input voltage is in phase with the input current. But notice that the output voltage swings negative as the input swings positive. We should have anticipated this phase shift from our previous discussions. The common emitter does have a 180-degree phase shift between input and output voltage. Notice also that the current curves in our area of operation (point D to point E) are perfectly linear. This means that our output waveshape is a faithful reproduction of the input; the output is free of distortion.

The power gain is equal to the current gain times the voltage gain. In this case, we have

$$G = A_i \times A_v$$

$$= 11 \times 88$$

$$= 968$$

Effect of Changing the Load With a given output bias, we can pivot the load line by changing the resistance of R_L. If we increase the resistance of R_L, the point of maximum I_c decreases. This lowers the left end of the load line and places it more nearly to a horizontal position. Decreasing the resistance of R_L increases the maximum I_c point. This action raises the left end of our load line and places it more nearly to a vertical position. Figure 5-8 illustrates this pivoting action.

Load line A is from our original circuit, which has an R_L of 4000 ohms. If we change the value of R_L to 8000 ohms, and keep the remainder of the circuit the same, our load line pivots from A to B. When we change the value of R_L to 2700 ohms, the load line pivots

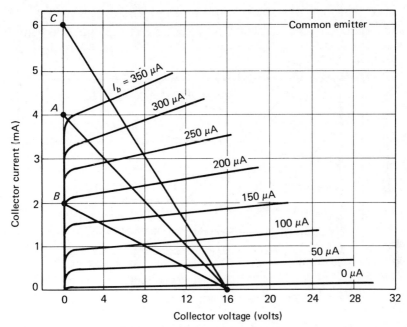
FIGURE 5-8. Effect of R_L on Load Line

to C. This pivoting action has great effect on the amplification of our circuit. This can be readily proved by using a given input signal and a given operating point, and plotting the output from each of the three load lines.

Effect of Changing the Collector Bias With a given value of R_L, the position of our load line is governed by the value of output bias. As we decrease the bias, the load line moves toward the lower left corner of the graph. When we increase the bias, the load line moves toward the upper right corner of the graph. We may plot a separate load line for any desired number of bias values, and all the load lines will be parallel to each other. Figure 5-9 shows a plot of three load lines for our original circuit with three different values of output bias. Our original load line is A. The same circuit with an output bias of 12 V produces load line B. When we change the output bias to 20 V, we have load line C.

If we plotted input and output signals on each of these three load lines, we would discover that gain is roughly the same in each case. We should conclude from this that the quantity of output bias has only a slight effect on the amplification qualities of a circuit.

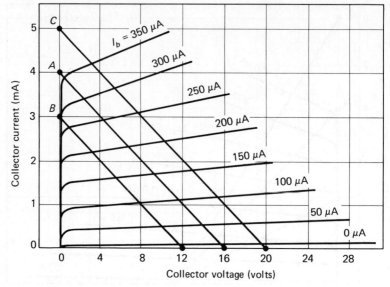

FIGURE 5-9. Effect of V_c on Load Line

DYNAMIC TRANSFER

With bipolar transistors, most of our references are to current because these are current-oriented devices. If we can determine the relative values of input and output currents, we can dispense with voltages for the most part. Once we have established a load line for a particular circuit, it is a simple matter to construct a second graph that shows a direct relationship between the values of input and output current. We call this current graph a dynamic transfer characteristic curve.

Constructing the Curve The idea is to take several points along the load line and project them onto an input-output current graph. The scale of the current graph is held the same as that used with the static curves and load line. We construct this scale by extending the grid lines to the left of our static characteristic graph. This procedure is clearly illustrated in Fig. 5-10.

The right-hand portion of this drawing is a family of static curves for a common emitter with a load line. We have placed a dot on each point of intersection of an I_b curve with the load line. These are the points we wish to project. We move the I_c designators to the left of our new grid and enter the I_b quantities along the bottom of

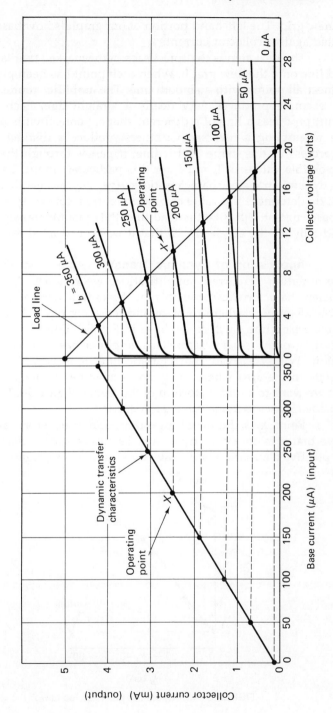

FIGURE 5-10. Projecting the Current Graph

the new grid. The left-hand portion of our graph is now base current plotted against collector current.

The next step is to project each of the marked points from the load line onto the new graph. When each point has been plotted, we connect all points with a smooth line. The amplifier represented by our example produces very nearly a straight line. Each different circuit produces a line of a different shape. Some circuits produce a line resembling a reversed C; others produce a distinct S shape. Regardless of the shape of this line, it passes through the locus of all possible values of I_b and I_c for this particular circuit. This line is our dynamic transfer characteristics curve. We can use this curve to quickly determine the value of collector current for any specific value of base current. But, like the load line, the dynamic transfer curve is valid only for the single circuit that it was constructed for.

Application of Dynamic Transfer Apart from the instant determination of current amplification, we can use the dynamic transfer characteristic curve to plot the output current waveshape and to determine where to place our operating point. With a linear characteristic line, such as we have in Fig. 5–10, the position of the operating point *is not* critical. This amplifier has practically no distortion. With a curved characteristic line, it is very important to operate in the area where the curve is most nearly linear, provided that we want to avoid distortion of the signal. Figure 5–11 shows an S-shaped curve and one cause of distortion.

Point X is the operating point, and it is set so low on the curve that it is on a curved portion. This causes some distortion on the positive alternation and heavy distortion on the negative alterna-

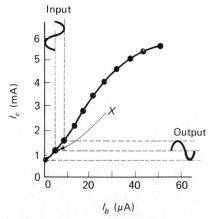

FIGURE 5–11. Operating Point Set Too Low

tion. Remember that the input bias and the value of R_b select our operating point. In this case, the static base current is much too low for proper amplification.

Distortion will also take place if our operating point is set too high on the curve. Figure 5-12 illustrates the results of an operating point set so high that it is on a nonlinear portion of the curve. The operating point is point X, and the output is badly distorted, especially the positive alternation.

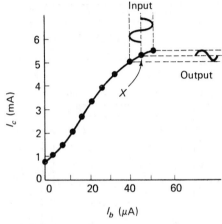

FIGURE 5-12. Operating Point Set Too High

Since we can control the location of the operating point by adjusting the input bias, we can set it to any desired position along our curve. If we are to avoid distortion, the best place for the operating point is at the center of the most linear portion of the dynamic curve, as shown in Fig. 5-13. Again, point X indicates the operating point, and now the output is a faithful reproduction of the input. With extremely large signals, the amplifier can still be driven into the nonlinear portion of the curve. When this happens, of course, distortion does result.

There are times when distortion in an amplifier is intentional. Remember that the class AB, B, and C amplifiers all produce alterations of the input signal. This is planned distortion, and this type of distortion can serve useful purposes. Suppose that we have a very large sine wave, and we wish to convert it to a square wave. A very easy way to make this conversion is to let the sine wave overdrive an amplifier. In other words, we use a class C amplifier with both cutoff and saturation limiting. Figure 5-14 illustrates this action with a dynamic transfer curve.

128 Large Signal Amplifiers

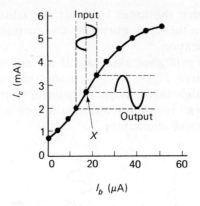

FIGURE 5–13. Operating Point Properly Positioned

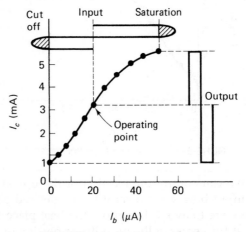

FIGURE 5–14. Overdriven Amplifier

This amplifier saturates when the base current reaches 50 μA, and it is considered cut off when the base current is zero. The large input drives the amplifier beyond both these extremes. Saturation limiting squares off the peak of the positive alternation, and cutoff limiting squares off the bottom of the negative alternation. The output is a fairly good square wave whose amplitude varies between 0.7 mA and 5.4 mA.

POWER DISSIPATION

All transistors have a power rating. This rating is in terms of the maximum continuous power that the transistor can safely dissipate. We must make sure that we do not require our transistor to dissipate power in excess of this rated value. There is a way graphically to depict the safe power so that we can be sure that the rating is never exceeded.

Constant Power Curve The constant power curve is a locus of all points where $I_c \times V_c =$ maximum continuous power rating. We are speaking in terms of a common emitter, and this power curve can be plotted on the V_c-I_c characteristics graph. Figure 5–15 is a graph of the constant power curve for a particular transistor.

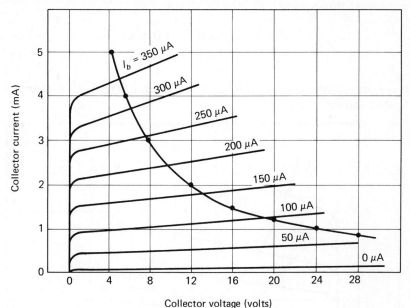

FIGURE 5–15. Constant Power Curve

This graph is for a transistor with a maximum continuous power rating of 24 mW at an ambient temperature of 25°C. The constant power curve passes through all points where $I_c \times V_c =$ 24 mW. We plotted the following points in order to draw the curve:

I_c	V_c
0.8 mA	30 V
1.0 mA	24 V
1.2 mA	20 V
1.5 mA	16 V
2.0 mA	12 V
2.4 mA	10 V
3.0 mA	8 V
4.0 mA	6 V
5.0 mA	4.8 V

After a sufficient number of points are plotted, we connect the points with a smooth curve. Our safe area of operation is clearly indicated; it is that area below and to the left of our constant power curve.

Relation to Load Line In practice we should construct the constant power curve before we construct the load line. Then we arrange our circuit values to produce a load line that never crosses the constant power curve. The load line for our previous circuit connected these points:

$$V_c = 16 \text{ V} \quad \text{and} \quad I_c = 0 \text{ mA}$$
$$V_c = 0 \text{ V} \quad \text{and} \quad I_c = 4 \text{ mA}$$

If we superimpose this load line on the graph of Fig. 5–15, it changes to that shown in Fig. 5–16.

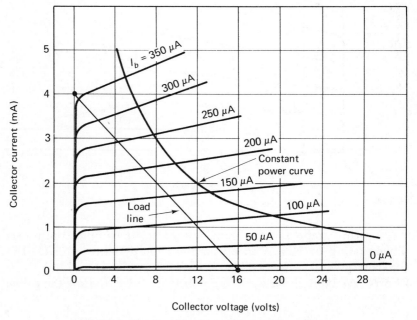

FIGURE 5–16. Combined Load Line and Power Curve

As you can see, we have a considerable distance between the two lines, even at the point of closest proximity. This distance is our margin of safety. With the exception of power amplifier circuits, we keep the load line and the constant power curve well separated. With power amplifiers, we are striving for maximum safe power. In this case, we may move the load line diagonally to the right until it touches the constant power curve. We can operate with a load line adjacent to the constant power curve, but the two lines must never cross.

Any point that falls above, or to the right of, the constant power curve overloads our amplifier. Using the graph in Fig. 5-16, let us pick such a point and calculate the power. Let us use $V_c = 8$ V and $I_c = 3.5$ mA. The power here is 3.5 mA $\times$ 8 V $=$ 28 mW. This point is only slightly beyond our curve, but we have exceeded our safe power rating by 4 mW. Any other point above the power curve will produce similar results. A surge of power that exceeds the power rating for a very brief period of time will probably do no harm, but a continuous excess power will certainly destroy the transistor.

FREQUENCY PROBLEMS

Transistors are fairly broadband devices, but we cannot expect any given transistor to handle all frequencies in the spectrum. Some transistors are, of course, designed to handle high frequencies. However, there is a point in the spectrum where each transistor is effectively cut off. Let us examine the reason for this frequency limitation.

Interelement Impedance We have discussed the input and output resistance of a transistor. However, impedance is a more appropriate term when one is dealing with ac, and impedance is composed of three factors: resistance, capacitance, and inductance. All three factors are always present to some extent, and the degree of prominence of a given factor is largely determined by the frequency. Since we amplify ac signals, each junction of the transistor becomes an impedance. There is also an impedance between the collector and emitter that manifests itself to a smaller degree. These three internal impedances are illustrated in Fig. 5-17.

At the frequency of operation, we can effectively ignore the effect of the inductance. This leaves us with impedances composed of resistance and capacitance as illustrated in Fig. 5-17. Each junction is a resistance in parallel with a capacitance as indicated for impedance A and B. Impedance C indicates the same type of arrangement between collector and emitter.

At low frequencies, the capacitance presents an open and has no effect on the signal. The impedance for these frequencies is

132 Large Signal Amplifiers

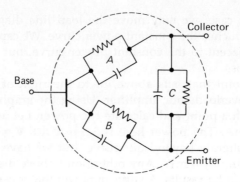

FIGURE 5-17. Internal Impedance

nearly pure resistance. As the frequency increases, capacitive reactance decreases. There comes a point in the frequency spectrum where the capacitor is a virtual short. Of course, effective amplification ceases at frequencies far below those that give us a direct short.

Transistors that are designed for high-frequency amplifiers are manufactured with very small interelement capacitances. Such careful attention to reduction of interelement capacitance greatly increases the capacitive reactance and moves the cutoff frequency farther up the spectrum. The interelement capacitance is also affected by the operating bias voltages. This is one reason why the manufacturer recommends a specific level of collector supply voltage. We should observe the recommended voltage fairly closely.

Internal Feedback The interelement impedance also causes a problem of internal feedback. This internal coupling makes our transistor a bilateral device; it couples signals in both directions. The problem becomes more pronounced as the frequency increases. If the feedback is in phase with the signal, it is regenerative and tends to cause the amplifier to oscillate. If the feedback is out of phase with the signal, it is degenerative and subtracts from the amplitude of the signal. The former condition causes too much amplification; the latter causes too little amplification. The oscillations caused from regenerative feedback will upset the stabilization of our amplifier. The attenuation caused by degenerative feedback results in a loss of power. We can construct an external feedback circuit that will cancel the effects of internal feedback. Such a circuit is illustrated in Fig. 5-18.

The values of R1 and C1 are chosen to match the values of the internal R and C across the base-collector junction. Signal A represents the signal that feeds back from the collector to the base

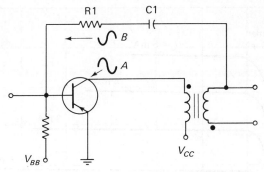

FIGURE 5-18. Neutralizing Circuit

internally. Signal B couples back through our external circuit. The two signals should be equal in amplitude and 180 degrees out of phase with each other. Thus, signal B exactly neutralizes the effect of signal A. When the effect of both internal impedance factors has been neutralized in this manner, our transistor becomes a unilateral device; effectively it couples signals only in the forward direction.

OTHER AMPLIFIER CIRCUITS

The bipolar transistor in a common emitter configuration is probably our most popular circuit for large signal amplifiers. However, other transistors and configurations are used. We shall use load lines and analyze some of these other circuits.

Common Base Amplifier We know that the common base configuration will have a current amplification that is less than unity. However, we do find that it is used because it is a good amplifier of both voltage and power. The static characteristics show input current I_e plotted against output voltage V_c and output current I_c. This graph is illustrated in Fig. 5-19.

A quick examination of the characteristics graph will reveal why this configuration has no current amplification. The emitter current is the input current, and it is always equal to the output current I_c plus the base current. When we change the input current, only 95 to 98 percent of that change can be registered as output current. We can check this on the graph. Let us hold the collector voltage constant at 10 V while we change I_e from zero to 4 mA. The 4-mA change in I_e causes I_c to change by about 3.8 mA. Alpha for this transistor is 0.95, and that is our current amplification.

134 Large Signal Amplifiers

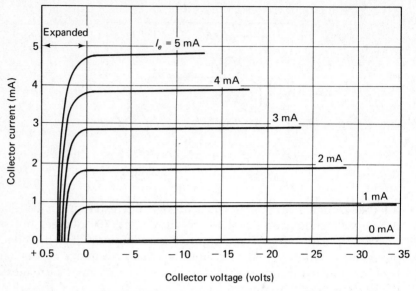

FIGURE 5-19. Common Base Characteristics

$$A_i = \frac{i_{\text{out}}}{i_{\text{in}}}$$

$$= \frac{3.8 \text{ mA}}{4 \text{ mA}}$$

$$= 0.95$$

Figure 5-20 is a schematic of a common base amplifier. We shall make use of the characteristics graph while analyzing this circuit.

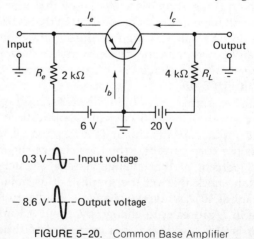

FIGURE 5-20. Common Base Amplifier

With circuit values as indicated, our static conditions are

$$V_c = -8.6 \text{ V}$$
$$V_e = 0.3 \text{ V}$$
$$I_e = 3 \text{ mA}$$
$$I_c = 2.85 \text{ mA}$$
$$I_b = 0.15 \text{ mA}$$

Our 0.3 V across the emitter-base junction V_e is based on the assumption that this junction has 100 ohms of resistance. This gives us 5.7 V across R_e and 0.3 V across the junction resistance. If you will recall the sample values for a common base configuration, input resistances range from 50 ohms to 150 ohms. Therefore, 100 ohms is a reasonable assumption. The emitter current is obtained like this:

$$I_e = \frac{V_e}{R_j}$$
$$= \frac{0.3 \text{ V}}{100 \Omega}$$
$$= 3 \text{ mA}$$

The emitter current is composed of I_b (5 percent) and I_c (95 percent). This gives us an I_c of 2.85 mA and an I_b of 0.15 mA. We calculate VR_L as follows:

$$VR_L = I_c \times R_c$$
$$= 2.85 \text{ mA} \times 4 \text{ k}\Omega$$
$$= 2.85 \times 10^{-3} \times 4 \times 10^3$$
$$= 2.85 \times 4$$
$$= -11.4 \text{ V}$$

V_c then becomes

$$V_c = V_{cc} - VR_L$$
$$= -20 \text{ V} - (-11.4 \text{ V})$$
$$= -20 + 11.4$$
$$= -8.6 \text{ V}$$

Therefore, our input signal will vary around 0.3 V and 3 mA. The output signal will vary around −8.6 V and 2.85 mA. When we construct a constant power curve and a load line for this amplifier, we have the graph in Fig. 5–21.

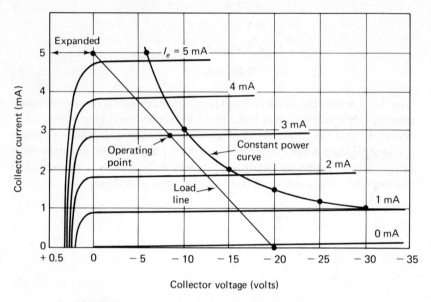

FIGURE 5–21. Common Base Load Line and Power Curve

The following coordinates were plotted to obtain our constant power curve:

I_c	V_c
1.0 mA	30 V
1.2 mA	25 V
1.5 mA	20 V
2.0 mA	15 V
3.0 mA	10 V
5.0 mA	6 V

With an output bias of 20 V and a load resistor of 4000 ohms, we have a load line from $V_c = -20$ V and $I_c = 0$ mA to $V_c = 0$ V and $I_c = 5$ mA. This places our amplifier in a safe area of operation powerwise, which is indicated by the fact that the load line is well away from the power curve. Our operating point is located where the load line crosses the static I_e line: $I_e = 3$ mA and $V_c = -8.6$ V.

Let us assume values for an input signal and analyze the results. Suppose that our signal is an ac sine wave, and its peak-to-

Other Amplifier Circuits   137

peak current is 2 mA. When we plot the input and output on our load line, we have the graph in Fig. 5-22.

Our total input signal is applied across the 100 ohms of input junction resistance. The signal must be 0.2 V peak to peak in order to produce the 2 mA change through the junction. As this signal swings positive, I_e swings upward from 3 mA to 4 mA. The increase in I_e causes I_c to increase from 2.85 mA to 3.8 mA. The increase in I_c causes more voltage drop across R_L and leaves a less negative potential for V_c. V_c swings from -8.6 V to -4.8 V; this is a change in the positive direction, because there is no phase inversion in the common base amplifier.

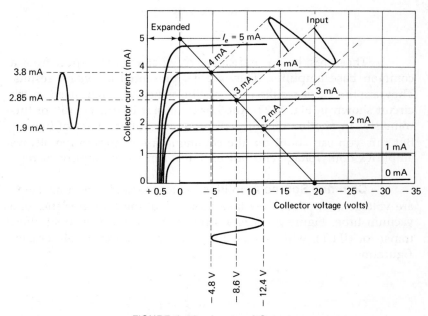

FIGURE 5-22.   Input and Output

On the negative half cycle of the input, I_e decreases from 3 mA to 2 mA. This decrease in I_e causes I_c to decrease from 2.85 mA to 1.9 mA. The decrease in I_c drops less voltage across R_L and V_c becomes more negative. V_c swings from -8.6 V to -12.4 V. We can calculate the current gain as follows:

$$A_i = \frac{\Delta I_c}{\Delta I_e}$$

$$= \frac{1.9 \text{ mA}}{2 \text{ mA}}$$

$$= 0.95$$

The voltage gain is

$$A_v = \frac{\Delta V_c}{\Delta V_e}$$

$$= \frac{7.6 \text{ V}}{0.2 \text{ V}}$$

$$= 38$$

The power gain is

$$G = A_i \times A_v$$

$$= 0.95 \times 38$$

$$= 37$$

These results are indicative of what we can expect from a common base amplifier. They *are not* presented as *typical* results. They are the exact results of a specific transistor when used in the circuit shown in Fig. 5–20. Another transistor in this circuit, or this transistor in another circuit, will produce a different set of figures. But this you can count on: the common base configuration always amplifies both voltage and power, but it can never amplify current.

Common Source Amplifier Since the field effect transistors are voltage-controlled, they take on many of the characteristics of a vacuum tube. Figure 5–23 is a schematic of a junction field effect transistor (JFET), which is used in a common source amplifier configuration.

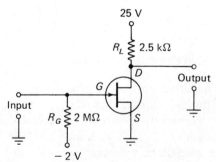

FIGURE 5–23. JFET Common Source Amplifier

This is an N-channel JFET, and in this configuration, we use a negative potential as bias across the gate-source junction. This bias must be of such a value that no part of the signal will cause the gate to become positive with respect to the source. If the gate is more positive than the source, gate current will result, and this creates intolerable distortion of the signal. The input bias in this case is set at -2 V. With the drain supply voltage as shown, and the 2500-ohm load resistor, our drain current can vary from 0 mA to 10 mA. The drain voltage has extremes of 0 V and 25 V. These values were used to plot the load line on the static characteristic graph in Fig. 5–24. The rated power of this JFET is 100 mW as represented by the constant power curve on the same graph.

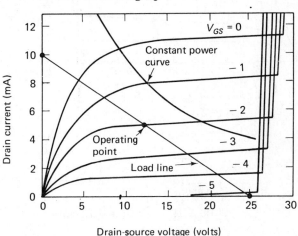

FIGURE 5–24. JFET Load Line and Power Curve

Our operating point is set by the point of intersection of the bias V_{GS} line and the load line. This quiescent point shows that we have a drain to source voltage of 12 V and a drain current of 5 mA. We shall use an ac sine wave of 2 V peak to peak as the input signal. The reaction of this particular amplifier to this signal is illustrated in Fig. 5–25.

The input signal rides our -2-V bias and swings the gate potential up to -1 V and down to -3 V. This causes the drain voltage to vary down to 7 V and up to 17.5 V. The positive alternation of the input signal is amplified slightly less than the negative alternation, but not enough less to cause any serious problem. The voltage amplification is

$$A_v = \frac{V_{out}}{V_{in}}$$

$$= \frac{-10.5 \text{ V}}{2 \text{ V}}$$

$$= -5.25$$

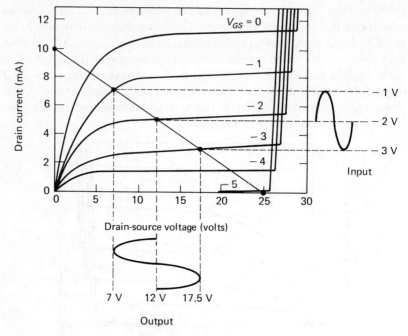

FIGURE 5-25. JFET Input and Output

The minus sign on the voltage amplification results indicates that we have a 180-degree phase inversion through this amplifier, just as we did through the common emitter.

The vertical lines on the graph indicate the avalanche breakdown point for various values of gate voltage. This JFET breaks down with a drain to source voltage between 25 and 30 volts for all values of gate voltage.

REVIEW EXERCISES

1. What is a large signal amplifier?
2. Where is the dividing line between large and small signal amplifiers?
3. What two factors determine the location of a load line?
4. What type of information can be obtained from a load line?
5. The load line for a particular common emitter amplifier is a straight line between what two points on the static characteristics graph?
6. Describe the operating point. How is it established?

7. Draw the circuit for a common base amplifier with an R_L of 4000 ohms and an output bias of 20 volts.
8. Assume that the transistor in the circuit of item 7 has an input resistance of 400 ohms and an input bias voltage of 4 volts. What value of R_b is required for an I_b of 200 µA?
9. Use the graph in Fig. 5–26 and construct a load line for the circuit of items 7 and 8.

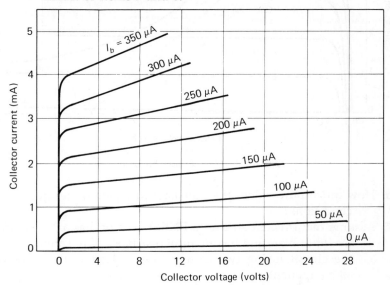

FIGURE 5–26. Static Characteristics

10. Refer to item 8 and locate the operating point in terms of I_b, I_c, and V_c. Mark this point on the graph of Fig. 5–26.
11. Use the load line just constructed to plot input current, output current, and output voltage for an ac sine wave input that is 200 µA peak to peak; indicate the values.
12. Refer to the amplifier of items 7 and 8 and calculate the
    a. Input voltage.
    b. Voltage gain.
    c. Current gain.
    d. Power gain.

NOTE: Items 13 through 21 may be answered by interpreting the graph in Fig. 5–27.

13. What type of transistor does this graph represent?
14. What type of circuit is represented?
15. What is the power rating of this transistor?

142 Large Signal Amplifiers

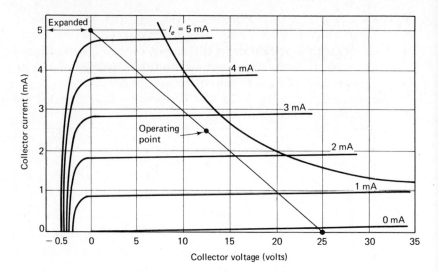

FIGURE 5-27. Interpret the Graph

16. How much power does the transistor dissipate under quiescent conditions?
17. Assume that this circuit uses a PNP transistor, has a 3-V supply for input bias, and has an input resistance of 100 ohms. Draw the circuit, using two-battery bias and show the value of all components.
18. What is the quiescent potential from emitter to base?
19. At the operating point, what is the potential at the collector?
20. Assume that we have an input signal of 3 mA peak to peak. What is the
    a. Peak-to-peak input voltage?
    b. Minimum and maximum output current?
    c. Minimum and maximum output voltage?
21. What is the gain of
    a. Current?
    b. Voltage?
    c. Power?

Chapter Six

Audio Amplifiers

Audio amplifiers are used to amplify signals in the audio range. The exact frequency limits of audio signals are a bit elusive. Some say 16 Hz to 20 kHz; others say 10 Hz to 20 kHz. At any rate, we can agree that they fall into the area of audible sound. For this reason, audio amplifiers are essential in all equipment which produces audible signals. This includes public address systems, radio receivers, television receivers, and sound recorders and reproducers.

The input circuit of an amplifier is powered by an input device. This input device may be an antenna, a signal generator, or simply the previous stage. Each amplifier is considered to be either a voltage, current, or power amplifier. It operates at a level higher than the previous stage and lower than the following stage. Audio amplifiers include preamplifiers, drivers, and power amplifiers. Preamplifiers usually operate at power levels measured in microwatts or even in picowatts. Driver stages are expected to operate in the milliwatt range, whereas power amplifiers may handle power in hundreds of milliwatts. Of course, these values vary with different types of equipment.

CURRENT AND VOLTAGE DISTRIBUTION

The current and voltage distribution are important factors in the analysis of any circuit. In this section we shall examine the actions of a transistor amplifier with respect to the current and voltage distribution in both input and output circuits. We shall first consider distribution during the quiescent state, then distribution with signals applied. We shall use a PNP bipolar transistor in a common emitter configuration for our discussion on distribution.

144   Audio Amplifiers

Quiescent State Current   All the conditions of a quiescent state are described by the operating point on a load line. These conditions exist when no signal is applied and repeat each time the signal value passes through zero. Figure 6-1 is a schematic of a transformer coupled audio amplifier. The currents indicated are quiescent currents caused solely from the bias voltages.

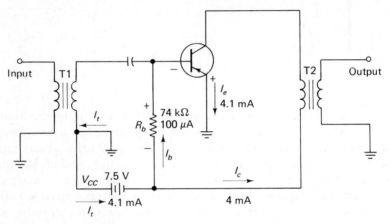

FIGURE 6-1.   Quiescent Current Distribution

The 7.5-V battery V_{CC} furnishes bias for both input and output. The input will be fed across T1 and the output will be taken from the secondary of T2. Electrons leaving the negative terminal of the battery compose the total circuit current I_t of 4.1 mA. This current divides at the first junction, splitting into I_b of 100 μA and I_c of 4 mA. I_b passes through R_b to the base, and I_c passes through the primary of T2 to the collector. I_b and I_c combine and emerge from the emitter as I_e of 4.1 mA. The emitter is connected to ground, and ground is connected to the positive terminal of the battery to complete the circuit.

When power is first applied, we have a momentary current flow into the right side of the coupling capacitor, but it quickly charges and blocks any further current. The primary of T2 also offers a momentary opposition to current on application of power. The first surge of current causes a field around the coil, but when current reaches a steady state, the coil acts as a short.

Quiescent State Voltage   Now let us examine the quiescent voltage distribution. Figure 6-2 is the same circuit with the voltage values indicated.

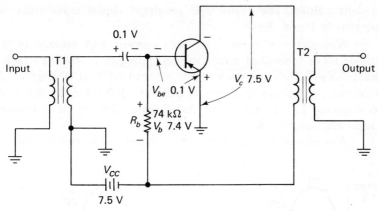

FIGURE 6-2. Quiescent State Voltage Distribution

Our current is in a steady state that leaves the coupling capacitor charged as indicated. The 100 μA of base current gives us a 7.4-V drop across R_b and leaves 0.1 V across the base-emitter junction with polarity as indicated. The charge on the coupling capacitor has to be the same as the voltage across this junction. We may now calculate the input resistance: the resistance of the base-emitter junction.

$$R_{in} = \frac{V_{be}}{I_b}$$
$$= \frac{0.1 \text{ V}}{100 \text{ μA}}$$
$$= \frac{0.1}{0.1 \times 10^{-3}}$$
$$= 1000 \text{ Ω}$$

Since we have no voltage drop across the primary of T2, the full value of V_{CC} appears on the collector. So V_c is 7.5 V negative with respect to the emitter. Thus we have the total applied voltage (7.5 V) dropped around both the input circuit and the output circuit. In fact, Kirchoff's voltage law can be applied to any complete loop that can be traced from the negative terminal back to the positive terminal of the battery.

Signal Waveforms Assume that we have an ac sine wave that is 200 μA peak to peak as an input signal to the circuit just discussed. The signal waveform will ride, and vary around, our quiescent values. The input and resultant signal waveforms are illustrated in Fig. 6-3.

Waveshape *a* represents both current and voltage of the input signal as measured across the secondary of T1. Notice that the signal reaches a positive peak of 100 µA and 0.1 V at 90 degrees. At 270 degrees, it reaches its negative peak of −100 µA and −0.1 V. At zero degrees, 180 degrees, and 360 degrees the signal is passing through zero amplitude.

Waveshape *b* represents both current and voltage for the

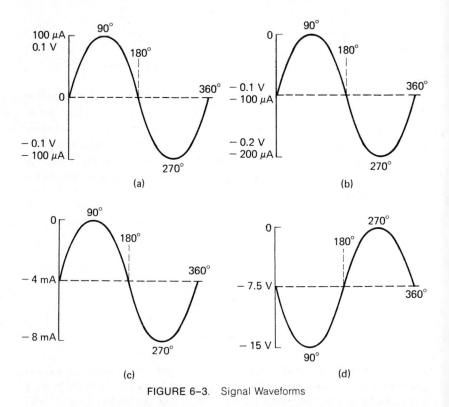

FIGURE 6-3. Signal Waveforms

signal as measured across the base-emitter junction. Notice that this waveshape rides the dc potential that we have across this junction. The signal here rises from −100 µA and −0.1 V to its positive peak of zero current and zero voltage at 90 degrees. It reaches its negative peak of −0.2 V and −200 µA at 270 degrees. At zero degrees, 180 degrees, and 360 degrees the signal is passing through the quiescent levels of current and voltage.

Waveshape *c* represents the collector current. This signal

current is riding the quiescent collector current of −4 mA. From −4 mA it rises to its positive peak of zero current at 90 degrees. The negative peak of −8 mA is reached at 270 degrees. At zero degrees, 180 degrees, and 360 degrees this signal is passing through the quiescent value of collector current.

Waveshape *d* represents the collector signal voltage. Notice that this waveshape is riding the quiescent collector voltage of −7.5 V. As the current rises, the collector voltage drops. It reaches its negative peak of −15 V at 90 degrees. The positive peak of zero volts is reached at 270 degrees. At zero degrees, 180 degrees, and 360 degrees this voltage is passing through the quiescent value of collector voltage.

Dynamic Current and Voltage For a true picture of the dynamic conditions, we shall consider the signal current and voltage at 90 degrees and again at 270 degrees. The conditions at 90 degrees are illustrated in Fig. 6–4.

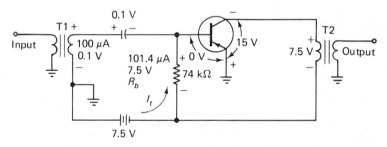

FIGURE 6–4. Signal Current and Voltage at 90 Degrees

At the 90-degree point, the signal on the secondary of T1 has reached its positive peak of 100 μA and 0.1 V. This 0.1 V is connected through ground to the emitter of the transistor. The signal voltage between emitter and base is equal in amplitude and opposite in polarity to the bias that we have across the base-emitter junction. This gives us zero volts between base and emitter, and this bias cuts off the transistor.

Another loop for the signal is through the battery, through R_b, and through the coupling capacitor. The coupling capacitor is charged as indicated and is large enough that the charge changes very little during signal time. Notice that this loop has three power supplies: the signal (0.1 V), the battery (7.5 V), and the charge on the capacitor (0.1 V). The signal voltage is equal in amplitude and opposite in polarity to the charge on the capacitor. These two voltages completely cancel, and this leaves 7.5 V to be dropped across

R_b. Dividing 7.5 V by the 74 kΩ of R_b gives us a current of 101.4 μA through this resistor. Since this is the only loop with current at this instant, this 101.4 μA is the total circuit current.

In the collector circuit, current has ceased because the transistor is now an open circuit. The quiescent I_c was —4 mA. (It was negative because its direction was toward ground through the emitter.) This current was decreased from 4 mA to zero. The change in I_c built up an opposing field across the primary of T2. At this instant, when I_c is zero, we have a charge across the T2 primary equal to the bias voltage of 7.5 V. The polarity of this voltage is as indicated, and it is in series, aiding with the battery potential. The voltage from collector to ground is now —15 V.

Keep in mind that the 90-degree point is not a static condition. We reached this point by sinesoidal changes, as depicted in Fig. 6–3. The changes reverse direction at this point, and the voltage and current pass through the quiescent conditions at 180 degrees. They continue in the same direction to 270 degrees. Let us freeze the action again at 270 degrees and examine the current and voltage distribution in the circuit. The conditions are indicated in Fig. 6–5.

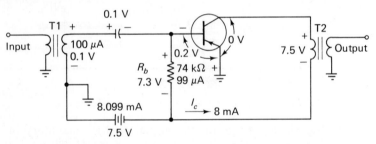

FIGURE 6–5. Signal Current and Voltage at 270 Degrees

When we reach the 270-degree point, our signal on the secondary of T1 has reached its negative peak of 100 μA and 0.1 V. The signal voltage, in series with the capacitor charge, places 0.2 V across the base-emitter junction with the base negative with respect to the emitter. Another path for the signal and capacitor current is downward through R_b and in series opposition to the battery. This opposing action across R_b cancels 0.2 V of the 7.5 V which was dropped here at the 90-degree point. This leaves us 7.3 V across the 74-kΩ resistor for a current of 99 μA.

The transistor is now acting as a short circuit with saturation collector current. The collector current is 8 mA, and it is negative because it is toward ground through the emitter. The collector current has built up an opposing field across the primary of T2 which is equal in amplitude and opposite in polarity to the battery voltage. The collector voltage at this instant is zero volts.

The total base current exists in the signal loop from the top of T1 secondary, through the capacitor, through the base-emitter junction to ground. We previously calculated the base-emitter resistance as 1000 ohms. The total base current at this time must be

$$I_b = \frac{V_{be}}{R_{be}}$$

$$= \frac{0.2 \text{ V}}{1000 \text{ }\Omega}$$

$$= 200 \text{ }\mu\text{A}$$

The changes reverse again at the 270-degree point, and the currents and voltages follow the sinusoidal pattern on to 360 degrees as indicated by the sine waves in Fig. 6–3. At 360 degrees the circuit is again passing through the quiescent conditions, and another cycle is underway. With this arrangement, we have obtained an output signal that is twice the amplitude of our output battery bias. This is possible because of the inductive action of transformer T2. With a stepup transformer, the output may be still higher.

PREAMPLIFIERS

Audio preamplifiers are low-level stages that usually follow low-level output devices such as transducers, microphones, hearing aid pickup, and recorder-reproducer heads. The primary considerations in preamplifiers are signal-to-noise ratio, input and output impedance, and frequency response.

Noise Factor The quality of an amplifier with respect to noise is indicated by a characteristic that we call the noise factor. We determine the noise factor by measuring the power signal-to-noise ratio at both input and output of the amplifier and dividing the input by the output ratio. If we designate the noise factor as F_o, the signal level as S, and the noise level as N, we can express this operation in an equation as follows:

$$F_o = \frac{S_{in}/N_{in}}{S_{out}/N_{out}}$$

Thus, the noise factor is equal to the input signal-to-noise ratio divided by the output signal-to-noise ratio. This equation indicates that the quality of our amplifier is inversely proportional to the noise factor.

The noise factor of an amplifier using a specific transistor is greatly affected by the quiescent conditions of the circuit. We set these conditions when we establish the load line and the operating point. As previously discussed, we set the operating point by selecting circuit values to give us a desired level of emitter current and collector voltage. The noise factor with respect to several values of emitter current and collector voltage is illustrated in Fig. 6-6.

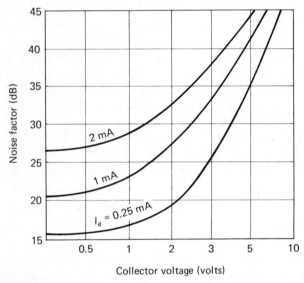

FIGURE 6-6. Noise Factor versus Emitter Current and Collector Voltage

This graph indicates that the noise factor for a given transistor is directly proportional to both the emitter current and the collector voltage. The best noise factor for this transistor is obtained by an operating point that gives us an emitter current less than 1 mA and a collector voltage less than 2 V. The graph also indicates that the noise factor increases more for an increase in collector voltage than it does for an increase in emitter current.

Our noise factor is also affected by the resistance of our signal source. We have said that the signal source may be considered as the previous stage, so our noise factor of an amplifier is affected by the output resistance of our previous stage. Figure 6-7 is a graph of noise factor as it is plotted against the source resistance.

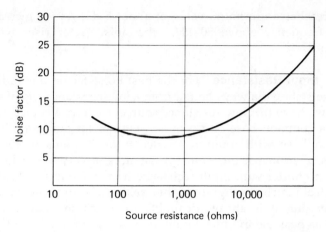

FIGURE 6-7. Noise Factor versus Signal Source Resistance

This graph shows that our noise factor decreases as the signal source resistance increases. This trend reverses when the resistance reaches about 800 ohms. From this point on, the noise factor increases as the source resistance increases. Our best noise factor for this transistor is obtained when the source resistance is kept between 100 and 3000 ohms.

The frequency of the signal has a considerable bearing on our noise factor. For very low frequencies, the noise factor is very high. As the frequency increases, the noise factor decreases for a time and then begins to increase again. Figure 6-8 is a graph of noise factor plotted against the frequency of the signal.

With this particular transistor, we have a noise factor of 30 dB when the signal frequency is 100 Hz. As the frequency increases, the noise factor slowly decreases until at 50 kHz the noise

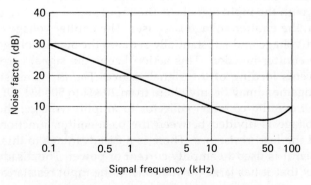

FIGURE 6-8. Noise Factor versus Frequency

factor reaches a minimum value of about 5 dB. As we increase the signal frequency above 50 kHz, the noise factor rises with the frequency.

Input Resistance For the best noise factor, our audio preamplifiers should always be fed from a low-resistance signal source. When we have this low-resistance source, we have a choice of using either a common base or a common emitter for our amplifier. The common base with input resistance from 50 ohms to 150 ohms enables us to match the source resistances of very low values. The common emitter with input resistance from 500 ohms to 1500 ohms is an ideal match for slightly higher values of source resistance. Remember that it is always desirable to match the impedances as closely as possible in order to obtain a maximum transfer of power.

Unfortunately, some of our signal sources have high resistances. When using such a source, we must work with what we have and contend with both high resistance and high noise factor. A crystal pickup head is one example of a high-resistance signal source. With this and other high-resistance sources, we must use a high-resistance input preamplifier in order to obtain a reasonable impedance match. Of course, we could use transformer coupling, and we do when the impedances are too far apart for other components. We have a choice of three high-resistance input arrangements without resorting to transformers. The common collector has a natural high input resistance. The common emitter can be changed to a high input resistance by connecting a resistor in series with either the base or the emitter. These three circuits are illustrated in Fig. 6–9.

The common collector amplifier illustrated in Fig. 6–9(a) has a high input resistance because of the large negative feedback voltage from emitter to base. The voltage across the load resistor is in phase with the input signal. As the input signal rises in a positive direction, the emitter voltage also rises. The emitter voltage opposes the input voltage and substantially reduces the signal voltage across the base-emitter junction. This action keeps the signal current low and prevents loading of the signal source. The input resistance of this arrangement may be anywhere from 20 kΩ to 500 kΩ. Of course, we *cannot* use the common collector to amplify voltage; the input signal voltage is divided between the base-emitter junction resistance and the load resistance. There is a disadvantage to this circuit, even when it is used to amplify current or power. The disadvantage is the fact that it has large fluctuations in the input resistance. Small

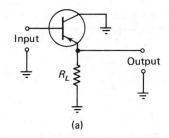

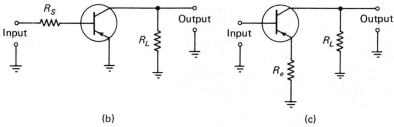

FIGURE 6-9. Amplifiers with High Input Resistance

changes in current drawn by the following stage will cause large changes in the input resistance of this stage.

In part (b) of Fig. 6-9, we have a common emitter amplifier with a resistor R_s in series with the base. The input signal circuit in this arrangement consists of R_s in series with the base-emitter junction resistance. The normal input resistance of 500 ohms to 1500 ohms is thus increased by the value of R_s. However, we must remember that the series resistance and the junction resistance form a voltage divider. For example, if $R_s = 1000$ ohms and $R_j = 1000$ ohms, only 50 percent of the signal appears across the junction. This series resistor can also cause serious instability problems, especially if it is part of the biasing circuit. On the plus side, the input resistance of this circuit will remain relatively constant even with large variations in the transistor parameters and current drain by the following stage.

The common emitter with degenerative feedback is illustrated in part (c) of Figure 6-9. Again our input signal circuit consists of the junction resistance in series with the emitter resistance R_e. The signal voltage developed across R_e opposes the signal voltage and reduces the gain of the stage. A big advantage of this arrangement is bias stabilization. The unbypassed emitter resistance acts as a swamping resistor that varies the bias in proportion to the current. Of course, this bias stability is not dependent on the negative feedback; we could bypass the resistor in the emitter and still have bias stabilization.

INTERSTAGE COUPLING

The frequency response of an amplifier, as well as the gain of the stage, is affected by the type of circuit that couples the signal between stages. We have four general types of circuits to choose from for this interstage coupling. These are RC networks, transformers, impedance networks, and direct wire. The advantages and disadvantages of each coupling circuit are determined by what we expect the circuit to accomplish. Some coupling circuits make use of variable components to adjust the gain of one or more stages. A gain adjustment in an audio amplifier is normally called a volume control.

RC Coupling We find that RC networks are widely used for interstage coupling. There are several reasons for this popularity. High gain, small size, and economy of circuit parts can all be accomplished with RC coupling. When we combine RC coupling with emitter swamping resistors and self bias, we can obtain a high degree of temperature stability. We may expect extensive use of RC coupling between stages of audio amplifiers in the following circumstances:

1. Low-level signal amplification.
2. Low noise preamplifiers.
3. High-level power stages.

Figure 6-10 illustrates two arrangements of RC coupling networks.

Figure 6-10(a) shows an arrangement of RC coupling which we call high frequency compensation. We are likely to see this coupling circuit on the output of an audio amplifier that uses a signal from a transducer with poor quality high frequency components. The signal is developed across R1 and coupled through C1. C1 has a relatively large capacitance, usually about 10 μF. A capacitance of 10 μF can be obtained in a capacitor of very small physical size because of the low voltage levels in this type of circuit. This large capacitance produces a very low capacitive reactance for all frequencies except the very low frequencies. For example, the reactance of a 10-μF capacitor to a frequency of 1000 Hz is

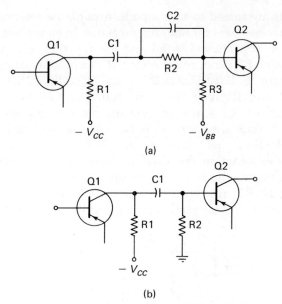

FIGURE 6-10. Examples of RC Coupling

$$X_C = \frac{1}{2\pi fC}$$

$$= \frac{1}{2 \times 3.14 \times 1000 \times 10 \times 10^{-6}}$$

$$= \frac{1 \times 10^3}{2 \times 3.14 \times 10}$$

$$= \frac{1000}{62.8}$$

$$= 16 \, \Omega$$

This same capacitor would offer a reactance of nearly 1600 ohms to a frequency of 1 Hz, so it effectively blocks the direct voltage that is used for bias.

R2 and C2 form an equalizer network to compensate for the fact that the low-frequency components of our signal are stronger than the high-frequency components. This network does not increase the amplitude of the higher frequencies; it simply offers more attenuation to the lower frequencies. This equalizes the high- and low-frequency components of our signal. Like C1, C2 offers less reactance to the higher frequencies than it does to the low. The low frequencies are forced to take a path through the resistor while the high frequencies pass through the capacitor. In either case, the signal input to Q2 appears across R3 and the junction resistance which are in parallel. R3 is not really essential except for biasing purposes.

Figure 6–10(b) shows RC coupling without the equalizing circuit. C1 and R1 have the same function that they had in the previous circuit. R2 is now our biasing resistor, and the full signal amplitude is impressed across the base-emitter junction.

Sometimes we have signals with predominant high-frequency components. In this case, we need low-frequency compensation to equalize the frequency components. This is accomplished by adding an RC network that will partially shunt the higher frequencies. Such an arrangement is illustrated in Fig. 6–11.

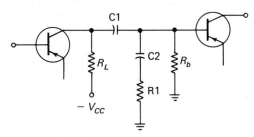

FIGURE 6–11. RC Coupling with Low-frequency Compensation

R1 and C2 compose the equalizing circuit. Both capacitance and resistance must be small to form an effective high frequency shunt that has no effect on the low frequencies. The capacitance is on the order of 1 or 2 μF, and the resistance ranges from 300 to 500 ohms. The capacitor offers a very high reactance to low frequencies and a very low reactance to high frequencies. The low-frequency components are coupled to the next stage with no alteration. The high frequencies are partially shunted, and the shunting effect increases with the frequency.

Transformer Coupling Transformer coupling is used primarily in situations that have a problem with impedance matching. The frequency response of transformer coupled stages is less than with RC coupling. Transformer coupling is also heavier, larger, and more expensive than RC coupling. Despite these disadvantages, transformer coupling will bring about maximum available power gain in a given amplifier. The high power efficiency and high output power are results of the close matching of the impedances. Transformer coupling is illustrated in Fig. 6–12.

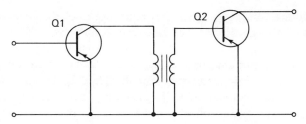

FIGURE 6–12. Transformer Coupling

The primary winding of the transformer is the collector load for Q1 and also the output coupling from the first stage. The secondary winding of the transformer serves as a dc return for I_b of Q2 and also as the input coupling for this stage. The very low resistance in the base of Q2 aids in temperature stabilization. This stability can be further improved by using a swamping resistor in the emitter circuit. If we neglect the natural transformer losses, which are very low, there are no power dissipating components in this coupling circuit. So, if our output impedance of Q1 matches the impedance of the primary, and the input impedance of Q2 matches the impedance of the secondary, we have a nearly perfect power transfer from Q1 to Q2.

Impedance Coupling An impedance coupling makes use of a coil, a capacitor, and a resistor, as illustrated in Fig. 6–13. We obtain high power efficiency with this type of coupling. The inductor is the collector load for Q1, and this provides high gain for all except the very low frequencies. At very low frequencies, the capacitor approaches an open as the coil approaches a short. As frequency increases, X_L increases and X_C decreases. Impedance coupling is effective for relatively high audio frequencies through the low radio frequencies. The frequency response of impedance coupled amplifiers is better than transformer coupling, but not as good as RC coupling.

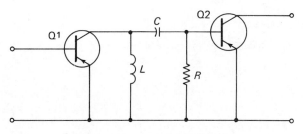

FIGURE 6–13. Impedance Coupling

158 Audio Amplifiers

Direct Coupling In direct coupling, we simply connect the output of one stage directly to the input of the next, as illustrated in Fig. 6-14.

This type of coupling is our only choice when we are coupling dc and very low frequency signals. Here we see the output of an NPN transistor connected to the input of a PNP. The base current of Q2 is the same as the collector current of Q1. If we have an I_c from

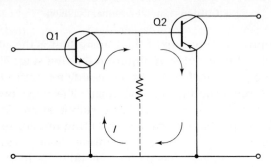

FIGURE 6-14. Direct Coupling

Q1 that is too large for Q2, we may connect a shunt resistor as indicated by the broken lines. However, connecting such a shunt will increase our power loss. The number of amplifier stages that we can directly couple is limited by the loss of stability. A current variation because of temperature in stage one will be amplified through all the stages. Figure 6-15 is a sample audio amplifier circuit using direct coupling.

This circuit was designed to compensate for a poor low fre-

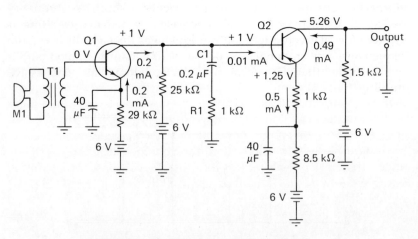

FIGURE 6-15. Direct-coupled Audio Amplifiers

quency output of the microphone (M1). A coupling capacitor between stages would further attenuate the already weak low frequencies. The direct coupling eliminates this undesirable event. The network formed by C1 and R1 equalizes the signals by shunting a portion of the high frequencies to ground. We call this action low-frequency compensation. The very low capacitance of C1 (0.2 μF) forms a sufficiently high-frequency short because of the high input impedance of Q2. This high impedance is primarily a result of the negative feedback caused by the unbypassed 1-kΩ resistor in Q2's emitter circuit.

The collector voltage and emitter current of Q1 are kept to a very low value to reduce the noise factor of this stage. Transformer T1 couples the microphone output to the base of Q1. The emitter of Q1 makes use of a bypassed swamping resistor for bias stabilization. The 40-μF capacitor shunts the ac signal around the resistor and prevents degenerative feedback. The signal from Q1 is developed across the 25-kΩ collector load resistor and is directly coupled to the base of Q2. The signal is amplified again through Q2, and the output is developed across Q2's 1.5-kΩ load resistor. The 8.5-kΩ resistor in the emitter circuit of Q2 is bypassed by another 40-μF capacitor. This circuit improves the bias stability of the stage.

CIRCUIT CONTROLS

We have two types of controls in audio circuits: volume controls and tone controls. A volume control varies the gain of one or more stages, whereas a tone control varies the frequency response. In transistor circuits, both of these controls are usually forms of current dividers with a variable resistor used as the activating component.

Volume Control A volume control should be so arranged that it introduces a minimum of noise into the circuit. This requirement is not as difficult as it sounds. Often we can achieve zero noise by avoiding heavy dc through the volume control. The volume control should also have a full range variation on the gain: from zero to maximum. Along with this action, the volume control must provide equal attenuation of all frequencies for all positions of the control. Figure 6–16 illustrates a volume control circuit that meets all these requirements.

The schematic in this figure is for two stages of audio amplification. The volume control R2 provides variation on the gain of the second stage. R1 is the collector load for Q1, and R6 provides

160 Audio Amplifiers

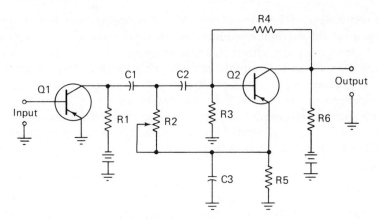

FIGURE 6-16. Volume Control Circuit

this function for Q2. R3 and R4 form a voltage divider to provide input bias for Q2. The bias on Q2 is stabilized by the swamping resistor R5, which is bypassed by C3. C1 and C2 are coupling capacitors, and both of them are vital to this circuit. Notice that these capacitors completely isolate the top of R2 from all sources of dc. This isolation prevents dc through the control and prevents noise generation.

Let us consider the signal path through R2. Capacitors C1 and C3 are virtual shorts for all signal frequencies. This places R2 in parallel with R1 as far as the signal is concerned. Therefore, any signal that is developed across R1 must also appear across R2. When we move the arm of R2 all the way to the top, the resistor is completely shorted. The shorting of R2 (minimum setting) provides practically a zero resistance path for the signal to go to ground. The path is through C1, around R2, and through C3 to ground. This is the setting for zero gain of Q2. As we move the arm of R2 downward, the resistance of this path increases. As the resistance increases, more of the signal is coupled into the base of Q2. When R2 is set to maximum resistance, maximum signal amplitude is coupled to Q2, and this is our maximum gain setting.

Tone Control The tone control circuit should control the frequency response as the volume control circuit controls the gain. Any frequency compensation circuit can be made into a tone control by including a variable element in the circuit. We use separate circuits for controlling low and high frequencies. The circuits in Fig. 6-17 illustrate both low-frequency tone control and high-frequency tone control.

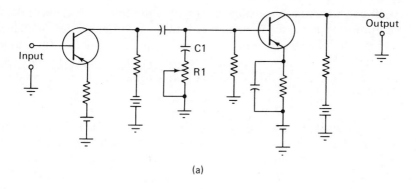

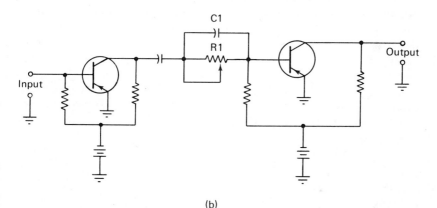

FIGURE 6–17. Tone Control Circuits

The tone control in both circuits is composed of C1 and R1. The circuit in part (a) of the figure uses a low-frequency compensation circuit with a variable resistor to provide a low-frequency boost. Part (b) of the figure uses a high-frequency compensation circuit with a variable resistor to provide a high-frequency boost. This high-frequency booster circuit is sometimes called a treble control.

AUDIO DRIVERS

Most of our audio power output stages use push-pull amplifiers and require two separate input signals. The two signals must be equal in amplitude and 180 degrees out of phase with each other. The phase requirements can be provided by a single-ended driver

162 Audio Amplifiers

with outputs from opposite ends of a tapped transformer secondary. A more popular way is to use a double-ended amplifier, which is commonly called a phase inverter or a paraphase amplifier. Of course, we can plan numerous ways to construct phase inverter circuits, but we shall limit this discussion to single-stage and two-stage inverters.

Single-stage Phase Inverter In most cases, we find that a single-stage phase inverter makes use of a common emitter amplifier with one output from the collector and another from the emitter. Such a circuit has considerable degeneration and may have a poor frequency response. If we have a strong input signal and can countenance low-frequency response, the single stage is, of course, the most economic. Figure 6–18 is a schematic for such an inverter.

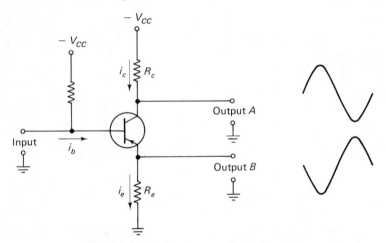

FIGURE 6–18. Single-stage Phase Inverter

The points on the schematic that are labeled $-V_{CC}$ are also ac grounds. All our dc power supplies are shunted by filter capacitors to provide these ac grounds. Then i_c is actually flowing from ground toward the transistor, and i_e is moving from the transistor toward ground. If i_c increases, more voltage is dropped across R_c, but this also increases i_e, which causes the voltage across R_e to swing more negative. So, output A and output B are 180 degrees out of phase with each other.

Since our emitter current i_e is composed of both i_c and i_b, it must be slightly more than i_c. Then if R_c and R_e are equal resistive values, output A will have slightly less amplitude than output B. It is generally desirable to have these two signals exactly equal in amplitude. We can equalize the output amplitudes in either of two ways: make the resistance of R_c slightly larger than that of R_e, or place a low-value resistor in series with the emitter output lead.

With strong signal currents, we may encounter a distortion problem because of unbalanced output impedances. The collector output impedance is higher than the emitter output impedance. This frequently happens in the single-stage inverter, because we generally use a large input signal. In fact, we must use a relatively large input because of the strong negative feedback from the unbypassed emitter resistor. Placing a resistor in the emitter output lead as indicated in Fig. 6–19 can balance the output impedances and eliminate the distortion. With careful selection of values for R1, R2, and R3, we can equalize the amplitude of the two signals at the same time that we eliminate the problem of distortion.

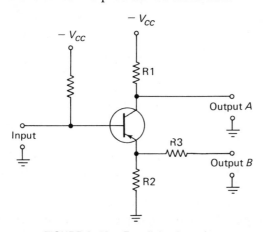

FIGURE 6–19. Equalizing Impedances

Two-stage Phase Inverters Inverters are driver stages for a double-ended amplifier and must provide a reasonably high power output. In many cases, we are unable to obtain sufficient driving power from a single-stage inverter. In Fig. 6–20, we use two common emitter amplifiers to form a two-stage inverter-driver.

This two-stage arrangement gives us full amplification of one stage, matched output impedances, and no distortion. This inverter provides adequate driving power for the push-pull power amplifiers. The signal through Q1 is amplified and inverted to provide both output A and the input to Q2. C_s is a coupling capacitor for coupling output A to the base of Q2. R_s is in the input signal path to decrease the amplitude of the Q2 input. The unbypassed resistor in the emitter circuit of Q2 provides negative feedback to reduce the gain and

164 Audio Amplifiers

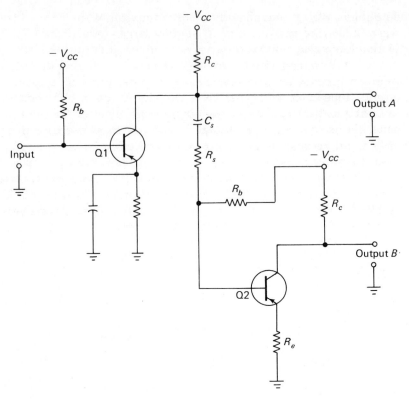

FIGURE 6–20. Two-stage Inverter

prevent distortion of the signal. The signal is inverted again through Q2 so that output B is 180 degrees out of phase with signal A. R_s and R_e are carefully chosen to provide just enough degeneration to counteract the gain of Q2. This degeneration gives us equal amplitudes for our two output signals.

POWER AMPLIFIERS

Single-ended amplifiers and push-pull class A amplifiers are frequently used in equipments that require neither a high power output nor a high degree of power efficiency. These amplifiers have the advantage of low distortion, which is a highly desirable characteristic of audio power amplifiers. However, the low output power generally prevents us from using them in our audio circuits. The class B push-pull amplifier provides higher output power and better power efficiency, and we can arrange the circuit to minimize the distortion.

Complementary symmetry is another arrangement that provides a very efficient power amplifier. We shall discuss the class B push-pull without bias and with bias, and then we shall consider some examples of complementary symmetry.

Push-pull without Bias For the sake of circuit simplicity we shall use transformer coupling for both input and output. We shall center-tap both transformers to provide our required phase difference. This circuit is illustrated in Fig. 6–21.

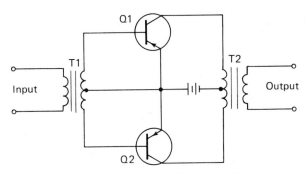

FIGURE 6–21. Push-pull with Zero Bias

The emitter-base junctions of both transistors are zero-biased. Each transistor conducts on alternate half cycles of the input signal. One alternation is amplified and inverted through Q1, and the other alternation is amplified and inverted through Q2. In the secondary of T2, the two half cycles are married to become a complete sine wave. This type of power amplifier generates intolerable distortion for an audio circuit. The reason for this distortion is illustrated in Fig. 6–22.

Part (a) of Figure 6–22 illustrates our dynamic transfer characteristics curve for a single transistor. When we place two identical transistors back to back in a push-pull circuit, this, in effect, places two characteristics curves end to end, as illustrated in part (b). Part (c) indicates the input and output current waveforms. The signal distortion is caused from the nonlinear portion of the curve, where one transistor very slowly stops conduction before the other very slowly starts to conduct. We call this crossover distortion, and it is a problem that must be eliminated. This crossover distortion becomes more severe with small input signals.

Push-pull with Low Bias We can easily eliminate the crossover distortion by placing a small forward bias across both base-

166 Audio Amplifiers

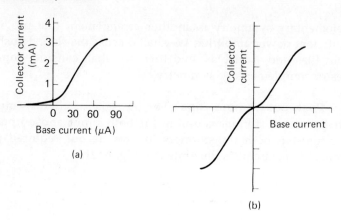

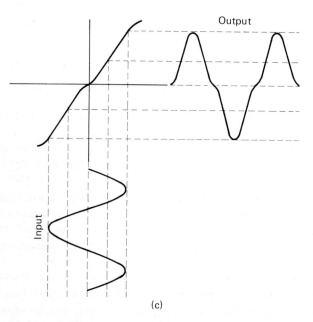

FIGURE 6-22. Zero Bias Dynamic Transfer

emitter junctions. This will allow both transistors to conduct for a short period of time just before and just after crossover. A simple method of providing this bias is illustrated in Fig. 6-23.

Resistors R1 and R2 form a voltage divider to keep the base of both our transistors negative with respect to the emitter. This gives us a small forward bias across both input junctions. Now each transistor will begin conducting slightly before the other cuts off. The effect of this bias on the transfer characteristics and the current waveform is illustrated in Fig. 6-24.

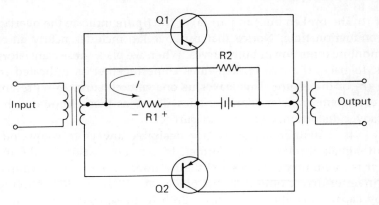

FIGURE 6-23. Forward-biased Junctions

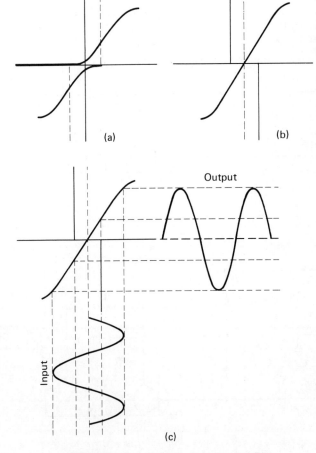

FIGURE 6-24. Dynamic Transfer with Bias

168 Audio Amplifiers

The broken lines in part (a) of the figure indicate the overlap in conduction time. Notice that this overlap includes nearly all of the nonlinear portion of both curves. When we place these transistors back to back, the two broken lines come together as indicated in part (b) of the figure. This leaves us one smooth curve to represent the composite transfer characteristics. The bias then eliminates the crossover distortion, as shown in part (c).

It is neither necessary nor desirable always to couple our input signals through a transformer. In fact, one reason for the inverter is to eliminate the necessity for a transformer. We can couple the inverter-driver output to our power amplifiers by RC coupling or by capacitance-diode coupling. Both of these coupling arrangements are illustrated in Fig. 6–25.

In part (a) of the figure we have RC coupling. Input A cor-

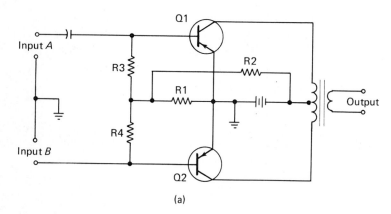

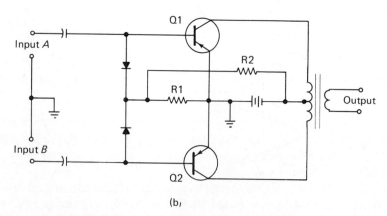

FIGURE 6-25. Coupling to Power Amplifier

responds to output A of our inverter. When this signal is positive, input B is negative and vice versa. When signal A goes positive, it places reverse bias on Q1 and cuts it off. At the same time, input B goes negative, increases forward bias on Q2, and causes it to conduct. On the next half cycle, Q2 is cut off while Q1 conducts. R3 and R4 are the base coupling resistors for Q1 and Q2, respectively. The function of R1 and R2 is the same as before; they develop dc bias for both transistors.

When we replace R3 and R4 with diodes, our circuit becomes that shown in part (b) of the figure. This is called capacitance-diode coupling. When a particular transistor base is positive, the associated diode conducts and shorts out the signal. When the base is negative, the diode is held cut off, and the input causes the associated transistor to conduct.

Complementary Symmetry All currents through a PNP transistor are exactly opposite in direction to the same currents through an NPN transistor. These current directions are illustrated in Fig. 6–26.

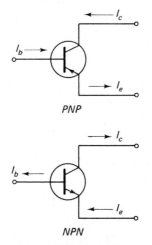

FIGURE 6–26. Direction of Currents

We can connect the two types of transistors into a single stage in such a manner that like currents complement one another. We call such an arrangement complementary symmetry, and it is illustrated in Fig. 6–27.

This is a circuit using two common collector configurations. With zero bias, as indicated in this drawing, current may or may not be present. But if currents are circulating, their direction will be

170   Audio Amplifiers

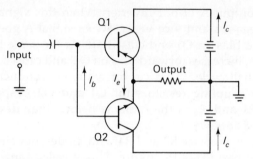

FIGURE 6-27. Complementary Symmetry (zero bias)

as indicated. When we add a proper biasing circuit to the schematic in Fig. 6-27, it becomes that in Fig. 6-28, which is a functional power amplifier.

Both transistors have forward bias, but only one is conducting at any given time. The portion of a signal that forward-biases one transistor reverse-biases the other. When our input signal swings negative, it forward-biases Q1 by forcing electrons from the capacitor through R3. This action reverse-biases Q2 and cuts it off. The current paths are i_c and i_b into the transistor and i_e out of the emitter to point A. From point A, total circuit current i_e goes through the output resistor to ground. The output during this alternation is negative with respect to ground, and this is the same polarity as the input signal. When our input signal swings positive, Q1 is reverse-biased to cutoff and Q2 is forward-biased to conduct. The current paths are now from ground through the output resistor to point A. This is both i_e for Q2 and total circuit current. This current develops an output signal that is the same polarity as our input: positive with respect to ground. The emitter current goes into the transistor and divides into i_b and i_c. The base current i_b goes upward through R3

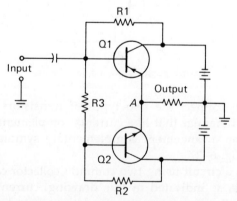

FIGURE 6-28. Properly Biased Complementary Symmetry Power Amplifier

and into the capacitor; from the capacitor it returns to ground through the signal source to complete the path.

The complementary symmetry circuit provides all the advantages of a push-pull amplifier, but it does not require a phase inverter driver stage nor a center-tapped transformer. The capacitor provides proper bias when charging, and when discharging, to greatly simplify the coupling circuit. The common load resistor with reversing polarities eliminates the need for an output transformer. We are not limited to the common collector amplifier when we use complementary symmetry. We can use the other configurations as well and can obtain more power gain, in some cases, if we do so.

REVIEW EXERCISES

1. What type of equipment uses audio amplifiers? List some examples.
2. What is meant by an input device to an audio stage?
3. List three divisions of audio amplifiers.
4. What does the operating point on a load line reveal about an amplifier?
5. What are the relationships of quiescent values and ac signal values?
6. What type of audio circuit is most concerned with signal-to-noise ratio?
7. How do we determine signal-to-noise ratio?
8. What is meant by the noise factor of a stage?
9. A given common emitter preamplifier has a proper operating point. What other factors affect its noise factor?
10. List two techniques of placing resistors that can be used to increase the input resistance of a stage.
11. List four types of interstage coupling commonly used in audio circuits.
12. What is a volume control?
13. Distinguish between a volume control and a tone control.
14. Draw and label the coupling circuits of item 11.
15. Draw an RC coupling circuit with high-frequency compensation.
16. Draw an RC coupling circuit with low-frequency compensation.
17. A low-frequency compensation circuit consists of a 0.2-μF capacitor in series with a 1000-ohm resistor. What is the X_C of this circuit to a frequency of:

    a. 1 Hz?
    b. 1 kHz?
    c. 20 kHz?

172 Audio Amplifiers

18. The output of a signal source is a low impedance and the input to the following stage is a high impedance. What type of coupling is best suited to this condition?
19. Why is impedance coupling not satisfactory for very low audio signals?
20. Explain why only direct coupling is suitable for dc.
21. Each of the coupling circuits in Fig. 6–29 contains a variable resistor. Identify the purpose of each.

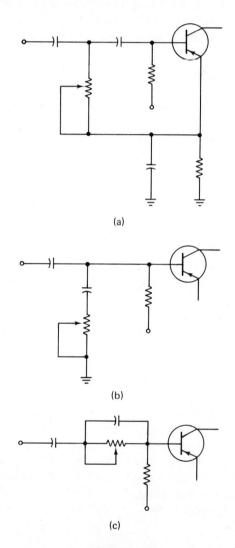

FIGURE 6–29.

22. Draw a schematic for a single-stage inverter-driver using a PNP transistor in a common emitter configuration.
23. What is the principal advantage of a two-stage paraphase amplifier over a single-stage inverter?
24. Why do we normally use class B, push-pull amplifiers for audio power amplifiers?
25. What causes crossover distortion? How can it be prevented?
26. Draw a schematic for a complementary symmetry power amplifier.

Chapter Seven

Tuned Amplifiers

Tuned amplifiers are narrow band amplifiers. They are tuned to a center frequency and amplify a narrow band of frequencies which extend to either side of this center point. The tuned amplifier is designed to pass and amplify this selected band of frequencies while suppressing all other frequencies. Selectivity is a term that describes how well an amplifier selects the desired band and discriminates against other frequencies.

INTERSTAGE COUPLING

The properties of our tuned amplifier depend primarily on the interstage coupling. For the most part, this coupling is composed of parallel resonant circuits. We rarely find a series resonant circuit used in this fashion, because it has no flexibility in impedance transformation and it forms a band reject filter rather than a band pass circuit. These parallel resonant tank circuits determine what frequencies will be passed and what frequencies will be rejected.

Another important feature of our interstage coupling is power transfer. Narrow band amplifiers are primarily power amplifiers, and we want a minimum power loss in the coupling circuits. To obtain this maximum power transfer, our coupling must match the output impedance of one stage and the input impedance of the next. The output impedance of a stage is usually very high while the input impedance is rather low. Our interstage coupling network must match these impedances while selecting the band of frequencies that we wish to amplify.

Characteristics of Parallel Resonance We are primarily interested in applications of parallel resonant circuits, but let us review some of their basic principles. Figure 7–1 illustrates a basic parallel resonant circuit and gives some of the key equations.

Our parallel tank circuit consists of capacitor C, inductor L, and the resistance of the inductor R_L. The first equation, $X_L = X_C$, expresses one condition of resonance. There are two other conditions that we consider to be resonant:

1. Line current in phase with the applied voltage.
2. Maximum current circulating in the tank.

Each of these three conditions may occur at a different value of frequency when we are dealing with a low Q circuit. As the Q of the circuit increases, these points come closer together. Since we shall deal only with very high Q circuits in this discussion, we may consider that the three conditions occur at the same frequency.

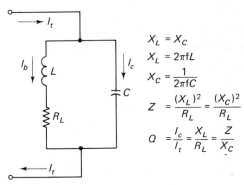

FIGURE 7–1. Parallel Resonant Circuit

The equations for X_L and X_C do not depend on resonance, but we derive our resonant frequency formula from the fact that these two quantities are equal when the resonant frequency is applied to the circuit. Since $X_L = X_C$, then

$$2\pi fL = \frac{1}{2\pi fC}$$

$$f^2 = \frac{1}{4\pi^2 LC}$$

$$f_r = \frac{1}{2\pi \sqrt{LC}}$$

We designate the resonant frequency as f_r to distinguish it from all other frequencies. It is the center of the band of frequencies that our tank circuit will pass.

The impedance of our tank at resonance is

$$Z = \frac{X_L X_C}{R_L}$$

And since $X_L = X_C$, we can simplify this equation as follows:

$$Z = \frac{(X_L)^2}{R_L} \quad \text{or} \quad Z = \frac{(X_C)^2}{R_L}$$

Impedance at resonance can also be calculated directly from the values of L, C, and R_L in this manner.

$$Z = \frac{L}{CR_L}$$

The impedance is very high at resonance and drops off for frequencies above and below f_r. The graph in Fig. 7-2 illustrates the impedance change in response to frequency change for both high and low Q circuits.

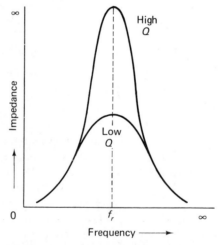

FIGURE 7-2. Impedance-frequency Curve

The width of the impedance-frequency curve is an indication of the resonant quality Q of the circuit; the Q is directly related to selectivity. A wide curve indicates a low Q or a poor quality circuit. As the Q increases, the curve becomes narrower. The Q is determined by the ratio of tank current (I_C or I_L) to current in the line (I_t). The ratio of X_L to R_L is also a measure of Q as is the ratio of tank impedance to either X_L or X_C. We can sum this up as follows:

$$Q = \frac{I_c}{I_t} = \frac{X_L}{R_L} = \frac{Z}{X_C}$$

The current that circulates in our tank circuit describes a curve very nearly like the impedance-frequency curve. The current peaks at resonance and drops off to both sides of resonance. The line current, of course, is just the reverse; it reaches a null at resonance and rises to either side of resonance. These currents are affected when we place a load on our resonant circuit, and this causes the Q to change. The effective Q of a circuit is the Q when the circuit is operating under normal loaded conditions.

A resonant circuit effectively passes the frequencies that fall between the low half-power point and the high half-power point. In a test situation, we can determine the frequencies at low and high half-power points by using a variable frequency signal generator and an oscilloscope. The frequency at the low half-power point will cause the current through the source resistor to lead the voltage across that resistor by 45 degrees. The circuit is capacitive when below resonance. Above resonance the circuit is inductive. The frequency at the high half-power point causes the current through the source resistor to lag the voltage across that resistor by 45 degrees.

We can also determine the effective limits of our bandwidth by finding the ratio between the resonant frequency and the effective Q of the circuit.

$$BW = \frac{f_r}{Q}$$

Problem: What is the resonant frequency, impedance at resonance, and Q of the circuit in Fig. 7–3?

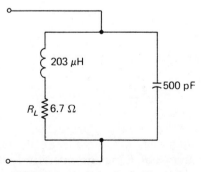

FIGURE 7–3. Solve for f_r, Z, and Q.

The resonant frequency is

$$f_r = \frac{1}{2\pi \sqrt{LC}}$$

$$= \frac{1}{6.28 \sqrt{2.03 \times 10^{-4} \times 5 \times 10^{-10}}}$$

$$= \frac{1}{6.28 \times 3.19 \times 10^{-7}}$$

$$= \frac{1 \times 10^7}{20}$$

$$= 500 \text{ kHz}$$

The impedance at resonance is

$$Z = \frac{L}{CR_L}$$

$$= \frac{203 \times 10^{-6}}{500 \times 10^{-12} \times 6.7}$$

$$= \frac{203 \times 10^4}{33.5}$$

$$= 60.6 \text{ k}\Omega$$

We need to find the value of either X_C or X_L in order to calculate the Q of the circuit. The inductive reactance is

$$X_L = 2\pi f L$$

$$= 6.28 \times 500 \times 10^3 \times 203 \times 10^{-6}$$

$$= 637.4 \text{ }\Omega$$

The quality of the circuit is

$$Q = \frac{X_L}{R_L}$$

$$= \frac{637.4 \text{ }\Omega}{6.7 \text{ }\Omega}$$

$$= 95.13$$

If we assume that R_L is the operating load on this circuit (and this will be true in some cases), we can calculate the bandwidth.

$$BW = \frac{f_r}{Q}$$

$$= \frac{500 \text{ kHz}}{95.13}$$

$$= 5.3 \text{ kHz}$$

The bandwidth is fairly narrow. The circuit is passing only the frequencies from 497.4 kHz to 505.3 kHz. If we double the resistance of R_L, we shall decrease the Q by 50 percent and double the bandwidth.

When we couple a parallel tank into a circuit, we normally place a resistive path in parallel with the tank, as illustrated in Fig. 7-4.

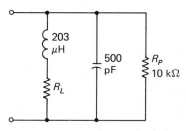

FIGURE 7-4. Parallel Loading

We have the same circuit as that shown in Fig. 7-3 with a resistance in parallel with the tank. We can ignore the small resistance of R_L in this case because the serious load on the tank is R_p. With this arrangement, we can determine the effective Q by assuming a voltage and calculating the currents. Let us assume an applied voltage of 10 volts for convenience. The applied voltage appears across each branch of the parallel circuit, but the tank current is independent of the line current. This tank current is either the current through the capacitor I_C or the current through the inductor I_L, and we determine its value as follows:

$$I_C = \frac{E_a}{X_C}$$

$$= \frac{10 \text{ V}}{637.4 \text{ }\Omega}$$

$$= 15.7 \text{ mA}$$

I_C and I_L are equal in amplitude and in opposite directions, and they

180 Tuned Amplifiers

effectively cancel each other. The line current I_t is then determined by the applied voltage across the parallel resistor. The line current is

$$I_t = \frac{E_a}{R_p}$$

$$= \frac{10 \text{ V}}{10 \text{ k}\Omega}$$

$$= 1 \text{ mA}$$

Our effective Q is

$$Q = \frac{I_C}{I_t}$$

$$= \frac{15.7 \text{ mA}}{1 \text{ mA}}$$

$$= 15.7$$

The bandwidth under these conditions has greatly increased with a corresponding reduction in selectivity. The bandwidth is

$$\text{BW} = \frac{f_r}{Q}$$

$$= \frac{500 \text{ kHz}}{15.7}$$

$$= 31.2 \text{ kHz}$$

Matching Impedances We may consider that the output of a transistor consists of a resistor R_o in parallel with a capacitor C_o. Also, the input of a transistor is composed of a resistor R_i in parallel with a capacitor C_i. These facts are illustrated in Fig. 7-5.

For maximum power transfer from transistor Q1 to transistor Q2, our coupling impedance between points 1 and 2 must equal the resistance R_o, and the impedance between points 3 and 4 must

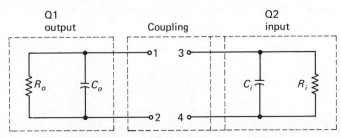

FIGURE 7-5. Output and Input Impedance

equal the resistance R_i. Normally the output capacitance C_o and input capacitance C_i are counted as part of our coupling network. If we need 500 pF of capacitance between points 1 and 2 and C_o is 10 pF, we use a 490-pF capacitor. The capacitances in parallel are additive, and this gives us our required total of 500 pF. We use the same technique to compensate for C_i between points 3 and 4.

Transformer Coupling In many applications we use transformer coupling with a tuned primary and an untuned secondary. A sample of this coupling circuit is illustrated in Fig. 7-6.

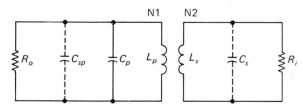

FIGURE 7-6. Tuned Primary to Untuned Secondary

Under these conditions the turns ratio of primary to secondary determines when our impedances are matched. We have a matched condition when

$$\frac{N1}{N2} = \sqrt{\frac{R_o}{R_i}}$$

The capacitance in the secondary of this circuit C_s is reflected back to the primary by transformer action to become C_{sp}. The value of this reflected capacitance is

$$C_{sp} = \frac{C_s}{N_r}$$

where N_r = turns ratio $(N1/N2)$.

Let us rearrange the circuit to that illustrated in Fig. 7-7.

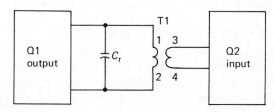

FIGURE 7-7. Simplified Tuned Primary Transformer Coupling

This is the same inductive coupling with tuned primary and untuned secondary. In this case, we are grouping all of the primary capacitance into C_t; this includes Q1's output capacitance and Q2's input capacitance that is reflected from the secondary. The output impedance of Q1 is matched to the input impedance of Q2 by selecting the proper turns ratio of primary to secondary. If we represent the primary inductance of T1 as L_p, then the resonant frequency of this circuit is

$$f_r = \frac{1}{2\pi \sqrt{L_p C_t}}$$

There are times when our calculated primary inductance is much too small for a practical transformer winding. We can alter the effective inductance, without changing the resonant frequency, by tapping the primary and adding capacitors to the primary. A tapped primary is illustrated in Fig. 7–8.

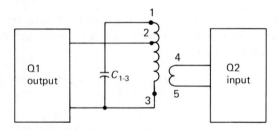

FIGURE 7–8. Tapped Primary

The inductance from 1 to 3 can be increased many times over our calculated inductance. To keep our frequency the same, we must add capacitors between 1 and 3 in a manner that total primary capacitance is reduced. The capacitance must be reduced by the same ratio as inductance is increased. If inductance from 1 to 3 is 100 times the calculated inductance, then the capacitance from 1 to 3 must be $\frac{1}{100}$ of the calculated capacitance C_t.

Autotransformer with Tuned Primary The autotransformer is another means of inductively coupling our signal between stages. The autotransformer has a single winding with taps to determine which portion of the winding serves as primary and which portion serves as the secondary. The basic coupling circuit is illustrated in Fig. 7–9.

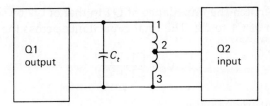

FIGURE 7-9. Autotransformer Coupling

The analysis of an autotransformer circuit is the same as for a conventional transformer using two windings. In this drawing, our primary consists of the turns between points 1 and 3. A portion of the same winding (points 2 to 3) composes our secondary. Capacitor C_t includes all the primary capacitance, both the physical and that reflected from the secondary. The primary inductance is all the inductance between 1 and 3, while the secondary inductance is the total inductance between 2 and 3. The resonant frequency is

$$f_r = \frac{1}{2\pi \sqrt{L_{1-3}C_t}}$$

We may position the tap at point 2 to match the input impedance of Q2. Also, we may find the inductance between 1 and 3 is too small to give us the desired frequency selectivity. In this case we may change the connection at point 1 to achieve an arrangement similar to that described in Fig. 7-8.

Capacitance Coupling As our frequency increases, it becomes more and more difficult to obtain maximum power transfer (unity coupling) by use of transformers. The secondary winding requires fewer turns as the frequency increases. This problem is particularly bothersome when using the common base amplifier with its very low input impedance. We can still have our tuned coupling by arranging output capacitors as illustrated in Fig. 7-10.

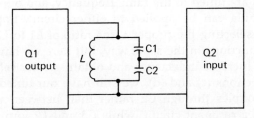

FIGURE 7-10. Capacitor Output Coupling

We match the impedance of Q1 to that of Q2 by selecting the proper ratio of C1 to C2. The total capacitance across the inductor is

$$C_t = \frac{C1\,C2}{C1 + C2}$$

The resonant frequency is

$$f_r = \frac{1}{2\pi\,\sqrt{LC_t}}$$

When Q2 has a very low input impedance, such as with a common base configuration, we may have difficulty with capacitor C2. The low input impedance effectively shorts this capacitor. In this case, we simply eliminate C2 from the circuit and leave that section of the circuit open. We may also connect an additional capacitor all the way across the coil with or without the use of C2. And we may tap the coil to increase the inductance as previously discussed in order to improve the selectivity of the circuit.

Double-tuned Networks We frequently find coupling circuits with a tuned output from one amplifier and a tuned input to the next. We call this double-tuned coupling, and it has several advantages over the single-tuned method. It provides a flatter frequency response between the half power points with a sharper cutoff below and above these points. In addition, it provides more attenuation for the frequencies outside of the bandpass for higher discrimination. The features just described make the double-tuned coupling a favorite for intermediate frequency amplifiers. Some double-tuned coupling networks are illustrated in Fig. 7-11.

In sections (a) and (b) of the figure, we have inductive coupling, and capacitive coupling is shown in sections (c) and (d). In part (a), C1 and L1 form a tuned circuit for the output of a stage, while C2 and L2 form a tuned circuit for the input to our next stage. Both circuits are tuned to the same frequency, and our resonant frequency formula can be applied to either circuit. Impedances are matched by selecting the proper turns ratio of L1 to L2. The circuit in part (b) functions in the same way, but here we have tapped the windings to increase inductance and obtain greater selectivity.

In sections (c) and (d), we still have our tuned circuits, but the signal couples through C2 rather than between windings. C1 and L1 form a resonant circuit, while C3 and L2 form another resonant circuit. The resonant frequency formula is applicable to either

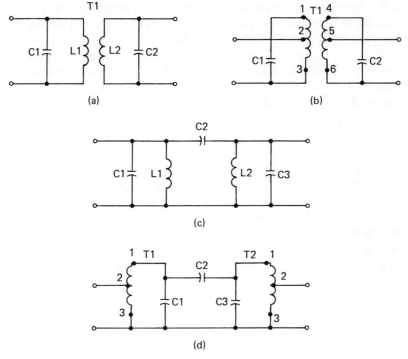

FIGURE 7-11. Double-tuned Networks

of these tuned circuits. In this case, we match the impedances by selecting the proper ratio of reactance between C2 and the impedance of the tank that is composed from L2 and C3. The circuit in (d) functions in the same manner with the added feature of tapped windings to enhance selectivity.

FEEDBACK ARRANGEMENTS

We have previously discussed the fact that a transistor *is not* a unilateral device; it couples signals in both directions. This internal feedback is regenerative (positive) in nature. It couples a portion of the output back to the input in a manner that aids the input signal. In audio frequencies, this is a small problem, but the problem grows as the frequency increases. If this internal feedback is strong enough, it will cause the amplifier to oscillate. High-frequency amplifiers with tuned coupling are susceptible to oscillations anyway, so we must take precautions to suppress these oscillations.

The effect of any type of feedback in a transistor circuit is a change in the input impedance. In most cases, feedback affects both the resistive and reactive components of our input impedance. We can offset the effects of internal feedback by countering it with external feedback. When our external feedback cancels both the resistive and reactive changes in our input circuit, our transistor is unilateralized. When our external feedback cancels only the reactive changes in the input circuit, the effects of our feedback have been neutralized. In either case, the external feedback prevents oscillations of the stage.

Unilateral Common Base Our common base amplifier has an internal feedback circuit that causes regenerative feedback. This circuit consists of the collector to base junction capacitance, the collector to base junction resistance, and the resistance of the base material. We refer to this base material resistance as the base spreading resistance. These feedback components and a circuit for unilateralizing the transistor are illustrated in Fig. 7–12.

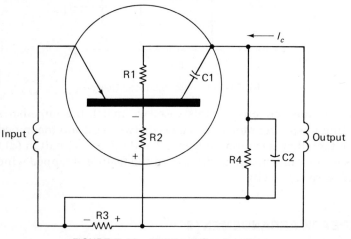

FIGURE 7–12. Unilateral Common Base

$R1$, $C1$, and $R2$ represent the internal feedback elements. $R1$ is the collector-base junction resistor. $C1$ is the collector-base junction capacitor. $R2$ is the base spreading resistance. When our input signal causes an increase in collector current I_c, a portion of this current couples through $R1$ and $C1$ to cause a voltage drop across $R2$ with the polarity as indicated. The polarity on $R2$ aids the incoming signal and constitutes a positive or regenerative feedback.

The external unilateralization circuit consists of $R4$, $C2$, and $R3$. The values of these components must correspond to the values of the internal circuit components: $R4 = R1$, $C2 = C1$, and $R3 = R2$. Now when our signal causes an increase in I_c, equal portions of this increase pass through $R2$ and $R3$. The polarities of the voltages on $R2$ and $R3$ are opposing and the amplitudes are equal. This completely cancels the effect of the internal feedback.

$R1$ and $R4$ are important components only with relatively low frequencies. At very high frequencies, $C1$ shorts out $R1$ and causes it to disappear from the circuit. The value of $R4$ must match that of $R1$ for the frequencies we expect to amplify, so at high frequencies we may be able to omit $R4$ from the circuit.

Common Emitter with Partial Degeneration With our common emitter circuit, we often use emitter degeneration as external degenerative feedback. We can unilateralize this configuration by returning a small portion of the emitter current to the collector through a feedback circuit. Such a circuit is illustrated in Fig. 7–13.

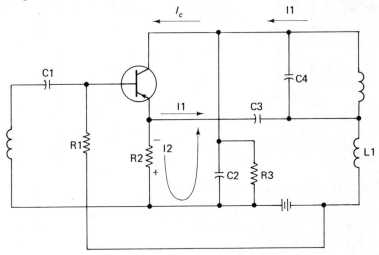

FIGURE 7–13. Unilateral Common Emitter

Resistor $R1$ provides forward bias for the input junction. Capacitor $C1$ blocks the dc and prevents shorting our battery through the transformer winding. Capacitor $C2$, in conjunction with output transformer primary, forms a parallel resonant circuit. Capacitor $C3$ couples a portion of the emitter current back to the transformer while isolating the top of $R2$ from the dc supply. The RF choke, $L1$, prevents an ac short across $R2$. The remaining components, $R2$, $C2$, and $R3$, compose our degenerative feedback circuit.

Notice that our collector current I_c is composed of $I1$ and $I2$.

When the input signal aids the forward bias, the collector current increases. This produces a positive feedback inside the transistor. The positive feedback is offset by directing a portion of I_e through R2 and back to the collector. This feedback current is labeled as I2 on the drawing. I2 develops a degenerating voltage across R2 that is equal and opposite to the internal feedback voltage. The values of R2, R3, and C2 depend upon the values of the base spreading resistance, the collector-base junction resistance, and the collector-base junction capacitance, respectively.

Bridge Feedback Circuit Unilateral feedback circuits frequently take the form of a bridge circuit, as illustrated in Fig. 7-14.

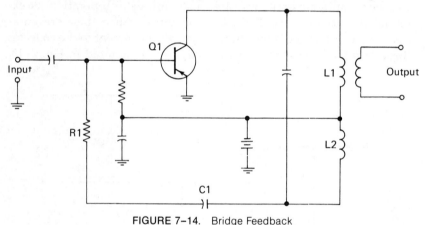

FIGURE 7-14. Bridge Feedback

The four legs of our bridge are

1. Collector to base impedance of Q1.
2. R1 in series with C1.
3. L2.
4. L1.

If we redraw the circuit, and leave out all components that are not part of the bridge, we arrive at a drawing similar to that in Fig. 7-15.

The voltage between points 1 and 3 is the voltage developed across coils L1 and L2. In a balanced condition, the voltage ratios on the four legs are

$$\frac{V_A}{V_B} = \frac{V_C}{V_D}$$

A balanced condition indicates that there is no internal feedback

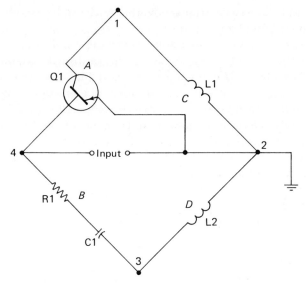

FIGURE 7-15. Simplified Unilateral Bridge

and that the voltage at point 2 is the same as that at point 4. An internal feedback upsets the balance and places a difference in potential between points 2 and 4. The difference in potential is external feedback with a polarity that opposes the input. This external feedback should exactly cancel the effects of the internal feedback.

AUTOMATIC GAIN CONTROL

We use tuned amplifiers extensively in radio and television receivers as RF and IF amplifiers. In these applications, we often find a need automatically to regulate the gain of an amplifier in accordance with the strength of its input signal. We want a high gain for a weak signal and a lower gain for a strong signal. One component of the output from a second detector is a dc directly proportional to the strength of the signal. We can use this dc component to vary the gain of a tuned amplifier. We may regulate the gain of an amplifier by controlling either the emitter direct current or the collector direct voltage.

Emitter Current Control As we increase the emitter dc of an amplifier from 0 to about 1 mA, we increase the power gain of the stage. As we increase the current above 1 mA, we decrease the gain, but this is a very slow decrease. The dc component of the detector output is called automatic gain control voltage, or simply AGC. The

190 Tuned Amplifiers

schematic in Fig. 7–16 illustrates how this AGC can be used to control the emitter current of an amplifier.

Here we have a tuned amplifier using a PNP transistor in a common emitter configuration. The input is directly coupled from the previous stage. The no-signal forward bias is established by R1 and R4, which form a voltage divider. The emitter is swamped by R3, which in turn is bypassed by C1 to prevent signal degeneration. The collector output signal is developed across the tuned circuit, which is composed of C2 and the transformer primary. The transformer is tapped to obtain the desired selectivity, and it matches the impedance to the next stage.

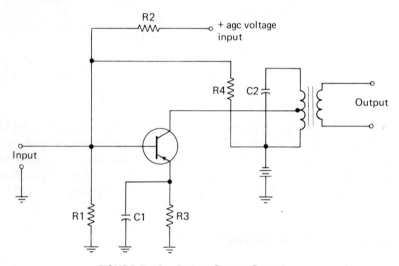

FIGURE 7–16. Emitter Current Control

The AGC voltage input to this amplifier is a positive direct voltage from the detector stage. The detector could be the next stage, or we could have several amplifier stages between here and the detector. This positive AGC voltage has a level that is directly proportional to the signal amplitude. The AGC voltage is coupled back from the detector, through dropping resistor R2, to the base of our transistor. Notice that the forward bias on the transistor is formed by a negative potential at the junction of R1, R2, and R4. This forward bias is directly opposed by the +AGC voltage. When our signal grows stronger, the +AGC voltage increases. This increased positive potential reduces the forward bias on our transistor and reduces the emitter current. This action reduces the gain of this stage and provides less amplification for the strong signal. When our signal grows weak, the +AGC voltage decreases and allows the forward bias on our transistor to increase. The larger forward bias causes an increase in emitter current, and this increases the gain of this stage. This action provides greater amplification for our weak signals.

Collector Voltage Control We may accomplish the same automatic gain control with a negative AGC voltage with a slight alteration on the stage. We feed the −AGC voltage to the same point in our amplifier, and it will, of course, vary the emitter current. But in this case, we shall add components to our amplifier stage so that we are exercising control of the collector voltage. This circuit arrangement is illustrated in Fig. 7–17.

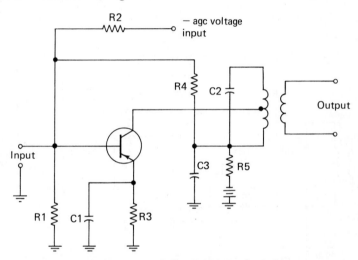

FIGURE 7–17. Collector Voltage Control

This circuit is identical to that shown in Fig. 7–16, with the exception of $C3$ and $R5$, which have been added to the collector circuit. The collector dc passes through $R5$, but the signal ac is bypassed by $C3$. $R5$ must be at least 10 kΩ to provide effective control.

The −AGC voltage on the base of our transistor aids the forward bias, which is provided by $R1$ and $R4$. When our signal amplitude increases, the −AGC voltage level increases (becomes more negative). The increased forward bias causes an increase in collector current and an increase in the voltage drop across $R5$. The increase in voltage across $R5$ reduces the voltage at the collector and reduces the gain of our stage. This action provides less gain for the strong signals. When our signal grows weak, the AGC voltage becomes less negative and reduces the forward bias on our tran-

192 Tuned Amplifiers

sistor. The reduced bias gives us a lower value of collector current. The lower collector current causes less voltage across $R5$ and a corresponding increase in collector voltage. With a higher value of collector voltage, our stage has more gain. This action provides more amplification for our weak signals.

Tuned Amplifier with AGC Let us consider a functional section of a radio receiver consisting of a double-tuned intermediate frequency (IF) amplifier, a second detector, and an AGC circuit. We shall use a PNP transistor in a common emitter configuration and control the gain by varying the emitter current. The schematic, with proper component values indicated, is shown in Fig. 7–18.

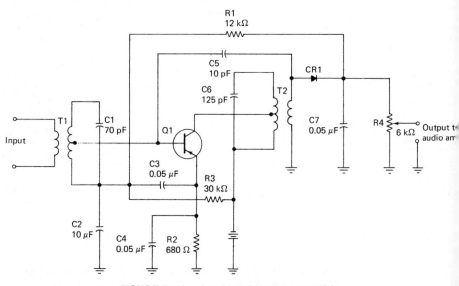

FIGURE 7–18. Amplifier, Detector, and AGC

The input to the primary of $T1$ is an IF carrier (center frequency 455 kHz), and it is amplitude modulated by an audio signal. Transformer $T1$ matches the output impedance of the previous IF amplifier to the input of this stage. The transformer also couples the input to the base of $Q1$. Capacitor $C1$ and the secondary of $T1$ form a parallel tuned circuit with a resonant frequency of 455 kHz. The secondary of $T1$ is tapped to provide the desired selectivity.

Transistor $Q1$ amplifies the input signal, which is developed across the tank circuit. The output of $Q1$ is developed across a second parallel resonant circuit consisting of capacitor $C6$ and the primary of $T2$. Transformer $T2$ also matches the output impedance of $Q1$ to the input impedance of the detector circuit and couples the modulated IF signal from $Q1$ to $CR1$. Resistor $R3$ provides a fixed forward bias by holding the base of $Q1$ negative with respect to the emitter. Resistor $R2$ provides emitter swamping, and the bypass capacitor $C4$ prevents signal degeneration. Capacitor $C5$ provides negative feedback to neutralize the internal positive feedback of $Q1$. $C3$ forms a low impedance path for the 455-kHz signal from the low end of $T1$ secondary to the emitter of $Q1$. $C2$ is an audio decoupling capacitor that prevents feedback of the audio signal through $R1$ to the base of $Q1$.

The detector circuit consists of $CR1$, $C7$, and $R4$. The modulated IF signal is rectified by $CR1$. $C7$ and $R4$ form an IF filter which removes the 455-kHz component of the signal. The remaining audio sine wave is coupled from the variable arm of $R4$ to our next stage, which is an audio amplifier.

At the junction of $CR1$, $C7$, and $R4$ we have a positive direct voltage with a level proportional to the signal amplitude. We couple this +AGC voltage across, dropping resistor $R1$ to vary the bias on $Q1$. This +AGC voltage cancels a portion of the forward (negative) bias provided by $R3$. The variation in bias controls the emitter current and regulates the amplification of $Q1$. The gain of $Q1$ is inversely proportional to the amplitude of the signal.

REVIEW EXERCISES

1. What is meant by a "tuned amplifier"?
2. Define the term "selectivity."
3. What portion of an amplifier's circuit provides the tuned properties?
4. What type of circuits is most often used for interstage coupling?
5. Why are series resonant circuits seldom found as interstage coupling in tuned amplifiers?
6. What characteristic of a coupling circuit provides maximum power transfer between two tuned amplifiers?
7. Name the three conditions that are considered as resonance in a parallel tank circuit.

194 Tuned Amplifiers

8. What type of tank circuit is likely to produce each of the three conditions of item 7 at a different value of frequency?
9. What is the relationship of the resonant frequency and the band of frequencies passed by a tank circuit?

Items 10 through 15 refer to the circuit in Fig. 7–19.

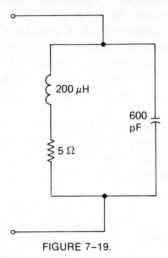

FIGURE 7–19.

10. What is the resonant frequency?
11. What is the tank impedance at resonance?
12. What is the Q of the circuit?
13. What is the bandwidth?
14. What frequencies are passed by the circuit?
15. If the resistance of the inductor is increased to 10 ohms, how will this affect the:

    a. Resonant frequency?
    b. Impedance at resonance?
    c. Q of the circuit?
    d. Bandwidth?
    e. Bandpass?
    f. Selectivity?

Items 16 through 20 refer to the circuit in Fig. 7–20.

16. What value of current is circulating in this tank?
17. What is the total line current?
18. What is the Q of this circuit?
19. What is the bandwidth?

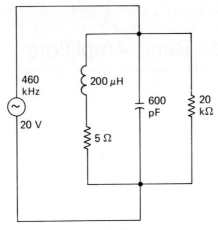

FIGURE 7-20.

20. If the 20-kΩ resistor is changed to a 40-kΩ resistor, how does this affect:
   a. Tank current?
   b. Line current?
   c. Q of the circuit?
   d. Bandwidth?
21. When interstage coupling between two amplifiers consists of a transformer with tuned primary and untuned secondary, how do we match the impedance?
22. Write the equation for a matched impedance between the two amplifiers of item 21.
23. What is an autotransformer?
24. List three advantages of double-tuned coupling over single-tuned coupling.
25. What type of feedback takes place inside a transistor?
26. What is the relationship between the signal frequency and the quantity of internal feedback?
27. How can the effects of internal feedback be canceled?
28. Why is it necessary to counteract the internal feedback?
29. What has been done to an amplifier circuit when the transistor is unilateralized?
30. When is a transistor neutralized?
31. What is the purpose of an automatic gain control?
32. How does an automatic gain control function?
33. List two popular methods of regulating the gain of an amplifier.

Chapter Eight

Wide Band Amplifiers

The amplifiers that we have previously discussed have been designed to amplify a narrow band of frequencies. They reacted best to one center frequency and discriminated against all frequencies except for a narrow band surrounding the center frequency. These narrow band amplifiers are used to amplify audio, intermediate, and radio frequencies, but they are not suitable for amplification of any type of nonsinusoidal signals. An examination of frequencies that compose a signal will emphasize the need for wide band amplifiers.

SIGNAL FREQUENCIES

When we speak of signal frequencies, we mean frequencies that are contained in a single cycle of the signal. We cannot point to the display of a sawtooth, a pulse, or a square wave signal and specify its frequency. We can specify how many times per second that the signal appears, but each cycle of these signals contains a large number of frequencies.

Sinusoidal Signals The signal that is shaped as a pure sine wave is the only signal that is composed of a single, fundamental frequency. This fundamental frequency may be different from one cycle to the next. For instance, when we are dealing with audio signals, the frequency may vary from about 15 Hz to about 20 kHz. However, the audio signal is a sine wave, and any given cycle of that train of sine waves contains only one exact frequency which we call the fundamental frequency.

To amplify these sine waves, we simply design amplifiers

with a bandpass wide enough to amplify the lowest and highest expected frequency. This includes a few kHz for signals of audio, intermediate, and radio frequencies.

Nonsinusoidal Signals Many areas of electronics are involved with nonsinusoidal signals commonly called video signals. We have the square-topped pulses of radar signals, stepped waveshapes of television video signals, and the sawtooth waveshapes of sweep signals. We must deal with sharp pulses, such as integrated waveshapes, and square waves as obtained from multivibrators. All these nonsinusoidal signals are composed of a fundamental frequency and many harmonics of this frequency. A harmonic is a multiple of the fundamental frequency. We consider the fundamental frequency ff as the first harmonic, $2 \times ff$ as the second harmonic, $3 \times ff$ as the third harmonic, and $50 \times ff$ as the fiftieth harmonic.

The stepped waveshape of television video signals is composed of frequencies from about 30 Hz to something in excess of 4 MHz. Remember that we are speaking of frequencies that are contained within a single cycle (or pulse) of the signal.

A perfect square wave is composed of a fundamental frequency and an infinite number of odd harmonics. Since we have no amplifiers that can handle an infinite band of frequencies, we cannot amplify a square wave without its suffering a certain amount of distortion. Consider the drawing in Fig. 8-1.

Here we see the start of a square wave. We have combined three sine waves and obtained something very different from a sine wave. As the harmonic frequency gets farther from the fundamental, the amplitude grows smaller, but each sine wave plays a part in the shape of the combination. Our resultant combined waveshape is exaggerated. We would need many odd harmonics to come so close to a square wave.

Signal Distortion When we amplify a nonsinusoidal signal, we subject it to two possible types of distortion: amplitude and phase. If one harmonic is amplified more than another, our signal suffers some form of amplitude distortion. If there is a phase shift, the shift must be proportional to the frequency; otherwise, our signal suffers from phase distortion. For example, if the fundamental frequency is shifted by 45 degrees, the second harmonic should be shifted by 90 degrees, the fourth harmonic should be shifted by 180 degrees, and so on, in order to maintain a constant phase relation among all harmonics. Figure 8-2 illustrates some of the distortions of a square wave.

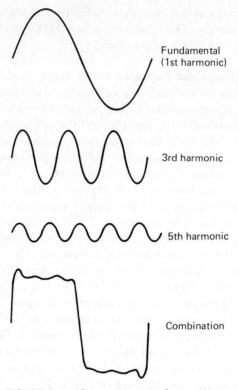

FIGURE 8-1. Composition of a Square Wave

In this drawing HF means high frequencies and LF means low frequencies. These are only a few of the possible distorted waveshapes that you may encounter, and these are intended only as rough approximations. This picture should serve to emphasize the importance of wide band amplifiers for all types of signals except pure sine waves.

TYPES OF COUPLING

As in tuned amplifiers, the frequency response of wide band amplifiers largely depends on the type of coupling that is used and the value of the coupling components. When dealing with a band of frequencies 4 MHz wide, we need a carefully designed coupling network for equal treatment of all frequencies that compose the signal.

Bifilar Transformers Some radar sets occasionally make use of a special transformer that has three windings. It is called a

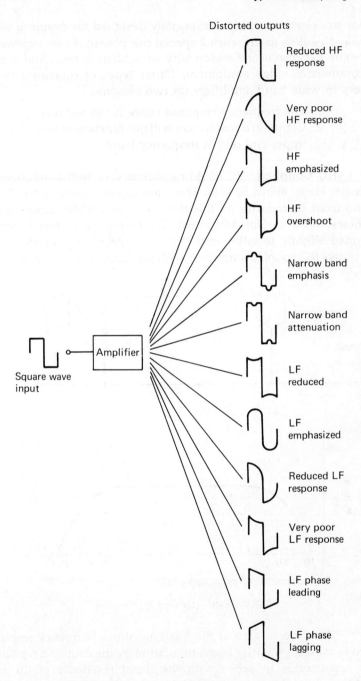

FIGURE 8-2. Phase and Amplitude Distortion

200 Wide Band Amplifiers

bifilar transformer, and it is especially designed for coupling video signals. Needless to say, such a special component is very expensive. The extra expense is one reason why we seldom, if ever, find it used in commercial video equipment. Other types of transformers are useless in wide band amplifiers for two reasons:

1. The frequency response curve is too narrow.
2. Common transformers will not function at the upper end of this frequency band.

RC Coupling RC coupling serves very well at frequencies from 100 Hz to about 50 kHz. These are sample figures only. There are no fixed boundaries for this flat response, middle region of frequencies. With careful design the flat frequency response can be extended slightly to either end of these approximate limits. Figure 8-3 shows RC-coupled amplifiers with the frequency response curve.

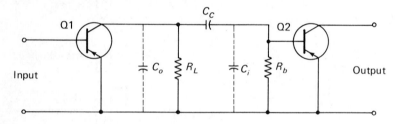

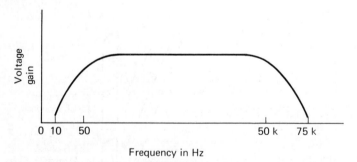

FIGURE 8-3. Common RC Coupling

The drop in gain at the low end of our frequency response curve is caused primarily from attenuation by the coupling capacitor. This capacitor is in series with the input resistance of the next amplifier. The signal is divided between X_C and R_j (junction resistance) in a ratio according to the values of these components.

Since X_C goes up as frequency goes down, the coupling capacitor exerts a severe attenuation on the low-frequency components of our video signal.

On the upper end of our response curve, we have a reduced gain because of the distributive capacitance C_o and C_i. As frequencies go higher, more and more of the signal is shorted to ground by these stray capacitances.

The value of the load resistor R_L also has a great bearing on our frequency response curve. Generally speaking, the higher the value of the load resistance, the more the amplification and the narrower the frequency response. A relatively low value of R_L tends to broaden the response at a cost of lowering the gain.

FREQUENCY RESPONSE

Since RC coupling seemed to be the only hope for wide band amplifiers, designers worked on ways to make it do the job. They had to find modifications to enhance both the high- and the low-frequency response.

Ideal Response The ideal frequency response curve for most wide band amplifiers is shown in Fig. 8–4.

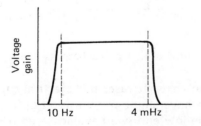

FIGURE 8–4. Ideal Response Curve

Notice that we have a perfectly uniform response to all frequencies from 10 Hz to 4 MHz. Also we have a sharp drop in amplification as soon as either limit is exceeded.

Practical Response While still striving toward this ideal, we find that wide band amplifiers do function reasonably well with a bandpass far from this ideal. Practically we find that most wide band amplifiers have a frequency response curve more nearly resembling one of the three shown in Fig. 8–5.

Wide Band Amplifiers

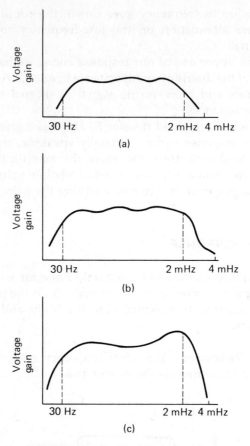

FIGURE 8-5. Practical Response Curves

Notice that in each case, the practical top of our band is about 2 MHz. We have slight amplification up to 4 MHz, but it is very slight. On the low end, we have achieved reasonable amplification down to 30 Hz, but that is just about the limit. Some of the irregularities along the top of these curves are caused from resonant conditions that were deliberately created to increase the bandwidth; others are caused from uncontrollable actions of the reactive components.

LOW-FREQUENCY RESPONSE

As previously stated, we have two separate problems in broadening the band. We must extend the response curve in both directions.

That means that the coupling must be modified in a manner that will increase its response to both low frequencies and high frequencies.

Low-frequency Attenuation At the low end of our response curve, the frequencies are attenuated by the coupling capacitor. The capacitive reactance for a particular capacitor increases as frequencies decrease. Since we are dealing with frequencies below 50 Hz on this end of the curve, the attenuation from X_C is very serious. Let us examine why this is so. Consider the drawings in Fig. 8–6.

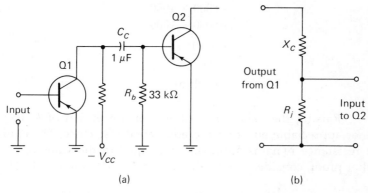

FIGURE 8–6. RC Coupling and Equivalent Circuit

In part (a) of the figure, we see an RC coupled amplifier. We have a coupling capacitor C_c of 1 μF and a base resistor R_b of 33 kΩ. But notice the equivalent circuit in part (b). Here the coupling is represented by the X_C of the coupling capacitor in series with the base-emitter junction resistance R_j. Why have we omitted R_b? Well, this is a common emitter configuration with input resistance R_j between 500 ohms and 1500 ohms. When the value of R_b is at least ten times that of R_j, R_b effectively drops from the picture. Let us use the drawing in Fig. 8–7 to clarify this aspect of input resistance.

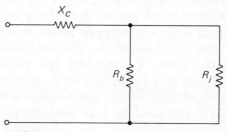

FIGURE 8–7. Coupling (input) Resistance

204 Wide Band Amplifiers

The base resistor R_b is in parallel with the junction resistor R_j. If R_b and R_j are equal in value, say 1500 ohms each, the equivalent resistance of the pair is half the value of either $\frac{1500}{2} = 750$ ohms). As the value of R_b is made larger, the equivalent resistor slowly increases. When the value of R_b is ten times the value of R_j, the equivalent resistance is approximately that of R_j. For example, when we have an R_b of 10 kΩ and an R_j of 1 kΩ, our equivalent resistance of the pair is

$$R_e = \frac{R_b R_j}{R_b + R_j}$$

$$= \frac{10{,}000 \times 1000}{10{,}000 + 1000}$$

$$= \frac{1 \times 10^4}{11}$$

$$= 909 \text{ ohms}$$

Raising the value of R_b above ten times that of R_j has no further appreciable effect on the equivalent resistance. The equivalent resistance remains approximately equal to the value of R_j. As further proof, consider these values: $R_b = 100$ kΩ and $R_j = 1$ kΩ.

$$R_e = \frac{100 \times 10^3 \times 1 \times 10^3}{100 \times 10^3 + 1 \times 10^3}$$

$$= \frac{100 \times 10^6}{101 \times 10^3}$$

$$= 0.99 \text{ k}\Omega$$

We increased the value of R_b ten times, from 10 kΩ to 100 kΩ, and the equivalent resistance changed less than 100 ohms. So for this discussion, we can forget about R_b and consider our coupling to consist of X_C in series with R_j, as depicted in Fig. 8–6.

X_C and R_j form a voltage divider for the signal voltage. A portion of the signal voltage is dropped across X_C and this portion is lost from our total signal amplitude. The remainder of our signal voltage is developed across R_j, and only this portion is actually coupled into Q2. Going back to Fig. 8–6, let us calculate some values to determine the seriousness of this low-frequency attenuation. At a frequency of 50 Hz, the X_C of the 1-μF capacitor is

$$X_C = \frac{1}{2\pi fC}$$

$$= \frac{0.159}{50 \times 1 \times 10^{-6}}$$

$$= \frac{159 \times 10^3}{50}$$

$$= 3180 \ \Omega$$

If our R_j has a value of 1000 ohms, the voltage ratio of X_C to R_j is about 3 to 1. This puts about 75 percent of the signal voltage across X_C and leaves only 25 percent across R_j. The 25 percent is the input to $Q2$, and the 75 percent is lost. As the frequencies go lower, the X_C gets larger, and the attenuation becomes worse. There are two approaches to improving low-frequency response: reduction of the attenuation and greater amplification of the low frequencies in the previous stage.

Decreasing Attenuation Since the coupling capacitor is causing the attenuation, one obvious step for improvement is to remove the capacitor. With direct coupling there is no capacitor, and all the signal is developed across R_j. This applies to frequencies all the way down to zero (dc). This method is sometimes used to solve the low-frequency response problem, but we cannot always use direct coupling.

The next best thing to eliminating the capacitor is to reduce its reactance to a minimum. We do this by making the capacitance larger. In our previous example, we found that a 1-μF capacitor at 50 Hz has a reactance of 3180 ohms. If we raise the capacitance to 10 μF, the reactance is

$$X_C = \frac{0.159}{50 \times 10 \times 10^{-6}}$$

$$= \frac{159 \times 10^3}{500}$$

$$= 318 \ \Omega$$

This capacitor in series with our 1000 ohms R_j would couple about 75 percent of our signal into the second transistor. That is satisfactory for 50 Hz, but how about 30 Hz? At 30 Hz, we still have a problem. The reactance is now

206 Wide Band Amplifiers

$$X_C = \frac{0.159}{30 \times 10 \times 10^{-6}}$$

$$= \frac{159 \times 10^3}{300}$$

$$= 530 \ \Omega$$

Now only two-thirds of the signal appears across R_j. We could increase capacitance some more, but there is a limit to this capacitance increase. For one thing, the capacitor's physical size increases with the capacitance. So if a capacitor must be used, the largest practical capacitance is selected. This larger capacitance keeps attenuation to a minimum, but greater amplification of the low frequencies is still necessary.

Increasing Amplification Increasing the amplification of the low frequencies will compensate for some of the attenuation that we still have in the capacitor. This is a simple circuit modification, and it is frequently used to improve low-frequency response. We add an RC component in series with the load resistor, as shown in Fig. 8–8.

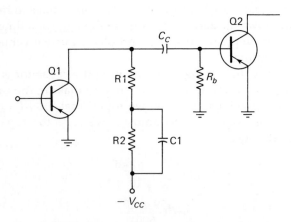

FIGURE 8–8. Low-frequency Compensation

$R1$ is our normal collector load resistor for $Q1$. $R2$ is an additional load in series with $R1$, but the capacitor $C1$ across this resistor makes the additional load a function of the signal frequency. At high frequencies, $C1$ is a short across $R2$ which removes $R2$ from the circuit. The total load for high frequencies then is the resistance of $R1$. As the frequencies decrease, the reactance of $C1$ increases and places more of $R2$'s resistance in the circuit. At very low frequencies, C1 is an open, and all the resistance of R2 is part of the collector load. With a greater load resistance, Q1 has a higher gain. So with this arrangement, we get a greater amplification of the low frequencies than we do for the high frequencies. Let us use some sample values to see how this works.

Suppose that R2 is 45 kΩ and C1 is 0.01 μF. The X_C of C1 needs to be ten times as large as the resistance of R2 in order effectively to place all of R2's resistance in the circuit. The X_C of a 0.01-μF capacitor at 30 Hz is

$$X_C = \frac{0.159}{30 \times 0.01 \times 10^{-6}}$$

$$= \frac{159 \times 10^3}{0.3}$$

$$= 500 \text{ k}\Omega$$

This is more than ten times the 45 kΩ of R2, so for all frequencies below 30 Hz, the full value of R2 is part of the collector load resistance.

HIGH-FREQUENCY RESPONSE

On the high end of our frequency response curve, we want to extend the response all the way from a few hundred kHz to, as near as possible, 4 MHz.

High-frequency Limitation Going all the way back to Fig. 8–3, we see that the two capacitors C_o and C_i are the main culprits in the limitation of the high frequencies. C_o is the output capacitance of one stage, and C_i is the input capacitance of the next stage. These *are not* physical capacitors. They represent the distributive capacitance between two wires, the capacitance across a transistor junction, and the capacitance between a conductor and a metal chassis. These capacitances have no effect on the low frequencies, but they definitely limit the high-frequency components of the signal.

Instead of an attenuation problem, we now have a shorting problem. The distributive capacitors are parallel to the signal path and are therefore parallel to the input resistance R_i of Q2. As our frequencies increase, the reactance of C_o and C_i decreases. With lower reactance, more and more of our signal is shorted to ground.

The total distributive capacitance may be grouped into one capacitance and placed in parallel with the collector load, as illustrated in Fig. 8–9.

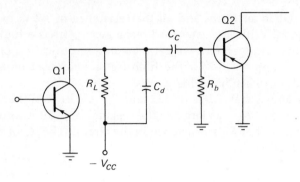

FIGURE 8-9. Distributive Capacitance

The distributive capacitance C_d is not very large. If we assume it to be 20 pF, we shall be pretty close. At 50 Hz this 20 pF offers a reactance of

$$X_C = \frac{0.159}{50 \times 20 \times 10^{-12}}$$

$$= \frac{159 \times 10^9}{1000}$$

$$= 159 \text{ M}\Omega$$

Obviously, with a reactance of 159 MΩ there is no shorting effect. But now consider this same capacitance at a frequency of 4 MHz. The X_C is now

$$X_C = \frac{0.159}{4 \times 10^6 \times 20 \times 10^{-12}}$$

$$= \frac{159 \times 10^3}{80}$$

$$= 2 \text{ k}\Omega$$

When this 2 kΩ is placed in parallel with the collector load resistor of $Q1$, about 2 kΩ, and the base-emitter junction of $Q2$, about 1 kΩ, it will shunt a large portion of the signal to ground. Since there is no way to dispose of the distributive capacitance, it must be put to use. It can be used in two ways: place coils in a manner to form it into either a parallel or a series resonant circuit.

Shunt Peaking One method of improving the high-frequency response is to place a coil in series with the load resistor of $Q1$. The coil is placed as illustrated in Fig. 8-10.

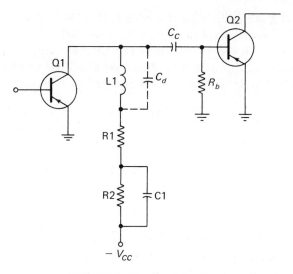

FIGURE 8-10. Shunt Peaking

The low-frequency booster circuit ($R1$ and $C1$) still appears in this amplifier circuit. This low-frequency circuit has no effect on the high frequencies. The coil $L1$ has been placed in series with the collector load resistor $R1$. The object of this arrangement is to form a parallel resonant circuit from the inductance of $L1$ and the distributive capacitance of C_d.

The resistance of $R1$ and the inductance of $L1$ are not arbitrary values. Best results are obtained when the resistance of $R1$ and the reactance of $L1$ are equal when the resonant frequency is applied. Since we are striving for a 4-MHz response, we can set our resonant frequency at 4 MHz. Now we turn our attention to the distributive capacitance. X_L and X_C will be equal at resonance, so if we calculate X_C at 4 MHz, we shall also have the value of X_L and $R1$. We previously assumed C_d to be 20 pF; let us use that value again. The reactance of 20 pF to 4 MHz is

$$X_C = \frac{0.159}{4 \times 10^6 \times 20 \times 10^{-12}}$$
$$= \frac{159 \times 10^3}{80}$$
$$= 2 \text{ k}\Omega$$

As you have probably noticed, we are rounding these figures off slightly. In both this and the previous example, the reactance was a little less than 2000 ohms, but the difference is so slight that it will

210 Wide Band Amplifiers

have no noticeable effect on the circuit. Then X_L at 4 MHz must also be 2 kΩ, and we will set the value of $R1$ to 2 kΩ.

Since coils are not purchased by their reactance, we still must determine the inductance that will cause this 2 kΩ of reactance at a frequency of 4 MHz. This is another mathematical process, as follows:

$$X_L = 2\pi f L$$

$$L = \frac{X_L}{2\pi f}$$

$$= \frac{2000}{6.28 \times 4 \times 10^6}$$

$$= 75.6\ \mu H$$

Again, we do not have to be exact. Tolerance allows us to round this off to the nearest convenient value and use a coil of a standard value.

Now we have an RC circuit composed of $R2$ and $C1$ which boosts the low frequencies and has no effect on the middle or high frequencies. In addition, we have a collector load $R1$ that handles signal development over the middle range of frequencies. Finally, we have added a circuit that will provide additional amplification for the high frequencies. In many video circuits, this may be considered adequate compensation to provide a sufficient bandwidth. However, we need to consider another broad banding technique.

Series Peaking Series peaking is a method of broadening the bandwidth by placing a coil in series with the signal path. It may be used alone or in conjunction with shunt peaking. Best results are obtained when series and shunt peaking are combined in the same circuit. However, in many cases, a small amount of quality is sacrificed in favor of reduced cost. The circuit in Fig. 8–11 shows the components in series peaking.

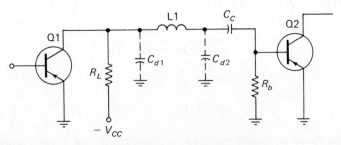

FIGURE 8–11. Series Peaking

The distributive capacitance is now divided into C_{d1} and C_{d2}. C_{d1} is roughly equivalent to C_o, while C_{d2} is roughly the same as C_i. The series coil $L1$ that has been added to our circuit divides these two capacitances. C_{d2} is now associated with $L1$, and we can choose an inductance for $L1$ that will form a series resonant circuit at the desired frequency. With the coil in the position indicated, the total distributive capacitance is about equally divided between C_{d1} and C_{d2}. If we continue to use our previous value of 20 pF for C_d, we have 10 pF for each of the capacitances. We then need an inductance that will resonate at our specified correction frequency when in series with 10 pF of capacitance. If we again assume our correction frequency to be 4 MHz, we may proceed as follows:

1. Calculate X_C of C_{d2} for 4 MHz.
2. Set $X_L = X_C$.
3. Use X_L formula and calculate L.

Then X_C is

$$X_C = \frac{0.159}{4 \times 10^6 \times 10 \times 10^{-12}}$$

$$= \frac{159 \times 10^3}{40}$$

$$= 3975 \; \Omega$$

Then X_L is also 3975 Ω, and

$$X_L = 2\pi f L$$

$$L = \frac{X_L}{2\pi f}$$

$$= \frac{3975}{6.28 \times 4 \times 10^6}$$

$$= 158 \; \mu H$$

A coil of approximately 158 μH placed as indicated will give us a series resonant circuit at a frequency of 4 MHz. The capacitive leg of this circuit begins to build up a very large voltage as the resonant frequency is approached. This boost in voltage adds to the amplitude of our signal.

The fact that our distributive capacitance has been split enables us to make other improvements in the circuit. We may now greatly enlarge the value of $Q1$'s collector load resistor for one very valuable consideration. We earlier calculated this optimum resistance to be equal to X_C of the total distributive capacitance, and we found it to be about 2000 ohms. But now only C_{d1} is in parallel with R_L, and we can raise the resistance of R_L to match the X_C of C_{d1}. The X_C of C_{d1} is the same as that of C_{d2}, which we just calculated: 3975 ohms. Then we can raise the value of R_L from 2000 ohms to about 4000 ohms. This increase in the value of the collector load resistance will increase the amplification over the entire frequency range.

We can also accomplish slightly different effects by changing the position of $L1$. If we place it between the collector of $Q1$ and the junction to R_L, we reduce the capacitance of C_{d1} and increase that of C_{d2}. Moving the coil so that it is placed between C_c and the base of $Q2$ accomplishes the opposite effect; it reduces the value of C_{d2} while increasing the value of C_{d1}. These variations in the placement of the coil enables the designer to divide the distributive capacitance into that which best suit his particular design.

When shunt and series peaking arrangements are compared, the series peaking provides the best broad banding results. However, as previously mentioned, the very best results are obtained by using both series and shunt peaking in the same circuit. When this is done, the two peaking circuits can be made to resonate at different frequencies and raise our frequency response beyond what either circuit can accomplish alone.

The Ringing Problem When a sudden voltage change occurs in a high Q resonant circuit, it has a tendency to cause undesirable oscillations of the tuned circuit. These oscillations are referred to as ringing, and they alter the shape of a video signal. The results of this ringing are illustrated in Fig. 8–12.

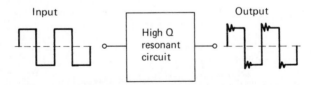

FIGURE 8–12. Ringing Distortion

The sudden changes of voltage on the square wave input cause overshoot and transient oscillations on the output from the resonant circuit. This ringing distortion is intolerable in video signals, and the peaking circuits must be designed to make sure that it does not happen.

Peaking coils may be purchased that have a built-in resistor in parallel with the coil. This resistor lowers the Q of the coil, broadens the response curve, and dampens the oscillations. Since these are all highly desirable qualities of high-frequency peaking circuits, we shall find this parallel resistor in most of these circuits. The schematic in Fig. 8–13 shows both series and shunt peaking with the damping resistors paralleling the coils. Low-frequency compensation is also shown in the same circuit.

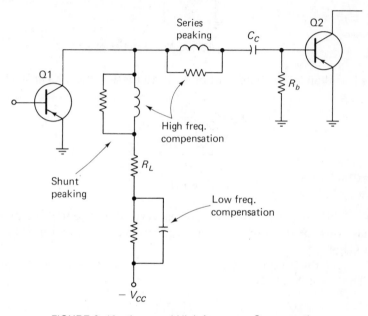

FIGURE 8–13. Low- and High-frequency Compensation

Enlarging the Input Impedance Another method of solving the low-frequency response problem is to raise the input impedance of the transistor. This is impossible with a given transistor, but we can exchange the bipolar transistor for a field effect transistor. Figure 8–14 is a schematic of a broad band amplifier using a junction field effect transistor (JFET).

This is an N-channel JFET in a common source configuration, and it has an input impedance on the order of 10 MΩ. This large input resistance enables us to place a resistor from gate to ground R_g of 1 MΩ with the input resistance being almost entirely determined by the value of R_g. A coupling capacitor of 0.1 μF, along with an R_g of 1 MΩ, provides a favorable division of voltage as long as

214 Wide Band Amplifiers

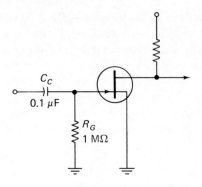

FIGURE 8-14. JFET as Wide Band Amplifier

X_C is less than 20 to 30 percent of R_g. With these values, X_C at 50 Hz is

$$X_C = \frac{0.159}{50 \times 0.1 \times 10^{-6}}$$

$$= \frac{159 \times 10^3}{5}$$

$$= 31.8 \text{ k}\Omega$$

This 32 kΩ in series with R_g causes very little attenuation of our signal. Most of the signal voltage appears across R_g as input to the next stage.

As the frequency decreases, the capacitive reactance goes up. The X_C at 20 Hz is

$$X_C = \frac{0.159}{20 \times 0.1 \times 10^{-6}}$$

$$= \frac{159 \times 10^3}{2}$$

$$= 79.5 \text{ k}\Omega$$

At this frequency, there is still only about 8 percent of the signal across C_c. With a frequency of 10 Hz, the signal will be attenuated by about 16 percent; this is still all right. At a frequency of 5 Hz, about 32 percent of the signal is lost across the coupling capacitor and leaves only 68 percent to be developed across R_g.

We should conclude that the use of a JFET as a wide band amplifier enables us to solve the low-frequency response problem by selecting the proper values of C_c and R_g.

The JFET *does not* solve the high-frequency problems. The

transistor itself has a better frequency response than a bipolar transistor, but the high-frequency peaking circuits are still required in many cases. The use of these circuits with the JFET is the same as described with the bipolar transistor.

SUMMARY

1. A nonsinusoidal signal such as a square wave, a sawtooth wave, and pulse waveshapes is composed of a fundamental frequency and many harmonics.
2. Wide band amplifiers are essential to a good reproduction of any nonsinusoidal (video) signal.
3. Video signals are subject to both phase and amplitude distortion.
4. Frequency response depends upon the type of coupling and the value of the coupling components.
5. The response to both low and high frequencies has to be improved in order to have a wide band amplifier.
6. Very low frequencies suffer from attenuation by the coupling capacitor. The X_C of this capacitor in series with R_j of the stage forms a voltage divider. The signal voltage across X_C is lost.
7. Low-frequency response can be improved by increasing the coupling capacitance, by adding an RC component in series with the collector load, and by increasing the input resistance of the next stage.
8. The input resistance can be raised greatly by using a JFET instead of a bipolar transistor for our amplifier.
9. With a JFET the voltage divider is C_c and R_g. Good low-frequency response can be obtained by selecting the proper values of these components.
10. High-frequency response is limited by the shorting effect of the distributive capacitance.
11. High-frequency response can be improved by either or both shunt peaking and series peaking.
12. Series peaking provides a better high-frequency response than shunt peaking; both series and shunt together do a better job than either type used alone.

REVIEW EXERCISES

1. Why is it impossible to use a narrow band amplifier with video signals?

Wide Band Amplifiers

2. What is the frequency composition of a
   a. Pure sine wave?
   b. Square wave?
3. In general terms, what is a proper bandwidth for signals of
   a. Audio frequency?
   b. Intermediate frequency?
   c. Radio frequency?
   d. Video frequency?
4. Name four types of signals that can be classified as video signals.
5. List two types of distortion that commonly occur with non-sinusoidal signals.
6. What is a bifilar transformer?
7. List two reasons why common transformers are not used in video amplifier coupling circuits.
8. How does common RC coupling in wide band amplifiers affect the
   a. Low frequencies?
   b. High frequencies?
9. What circuit components are chiefly responsible for the effects of item 8?
10. With a common emitter amplifier, how does the base-emitter junction affect the signal amplitude?
11. Describe the relationship of amplification and frequency response to the value of the load resistor.
12. A common emitter amplifier has an R_b of 20 kΩ. How is the low-frequency response affected by raising the value of R_b to 50 kΩ? Explain.
13. What is the equivalent input resistance of a common emitter that has an R_b of 5 kΩ and an R_j of 1500 ohms?
14. In item 13, what is the input resistance if R_b is raised to
    a. 15 kΩ?
    b. 50 kΩ?
15. A 1-μF coupling capacitor offers how much impedance to a
    a. 30-Hz signal?
    b. 20-Hz signal?
16. Consider the schematic in Fig. 8–15. Approximately what percentage of the signal amplitude is coupled into $Q2$?

17. Redraw the schematic in Fig. 8–15 and add an RC circuit in a manner that it will improve the low-frequency response.

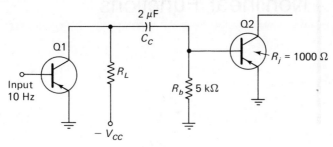

FIGURE 8–15.

18. Explain how the RC circuit (added in item 17) functions.
19. Draw an RC coupled, common emitter amplifier and indicate the primary hindrance to high frequencies.
20. Explain why the distributive capacitors of item 19 are detrimental to high frequencies.
21. Draw an RC coupled, common emitter amplifier with high-frequency resonant shunt peaking.
22. Suppose that the amplifier in item 21 has 25 pF of distributive capacitance and that we wish to have resonance at a frequency of 3.5 MHz. Calculate the proper values of R_L, X_L, and L for this circuit. Show your work.
23. Draw an RC coupled, common emitter amplifier with high-frequency resonant series peaking. Show the division of the distributive capacitance.
24. Assume that C_{d2} of item 23 is 15 pF. Calculate the X_L, X_C, and L for a resonant frequency of 3.5 MHz. Show your work.
25. What can be done to a high-frequency peaking circuit to prevent ringing distortion of the signal?

Chapter Nine

Nonlinear Functions

Many electronic devices have inherent nonlinear characteristics that lend themselves to a variety of useful applications. Perhaps the best-known of these functions are the logarithmic and the square law. We find that designers make a great variety of uses of these and other nonlinear functions. Amplitude compression, amplitude expansion, amplitude limiting, frequency response, multiplication, division, squaring, square rooting, and many other circuit types utilize nonlinear functions.

NONLINEAR DEVICES

Nearly all solid state devices exhibit some form of nonlinearity that can be turned to advantage through proper circuit design. These devices have nonlinearities in their electrical resistance, or in their transfer characteristics, which we can alter by a control signal. Let us examine the nonlinear characteristics of a few common solid state devices.

PN Junction Diode The PN junction diode has nonlinear forward characteristics that are logarithmic in nature. This is especially true of the silicon diode, where a change in diode voltage is a direct logarithm of the diode current. This logarithmic relationship of voltage and current holds true over a wide range of diode current. Mathematically we may express this relationship in terms of either current or voltage. In terms of current, we have

$$I = I_o \exp\left(\frac{q}{mkT}\right) V$$

where I_o = reverse saturation current, $q = 4.77 \times 10^{-10}$ (which is electron charge in esu), m = the effects of diffusion current, $k = 1.37 \times 10^{-15}$ (Boltzmann's constant), T = temperature of the junction (absolute), and V = voltage across the diode. In terms of voltage, the same relationship is expressed as

$$V = mkT (\log_{10} I - \log_{10} I_o)$$

where k is a constant. This logarithmic relationship of current and voltage may also be expressed graphically as shown in Fig. 9-1.

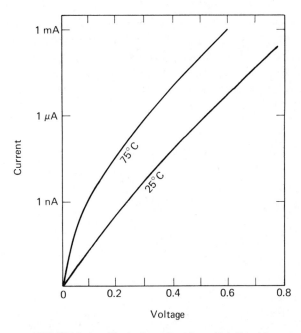

FIGURE 9-1. Diode Current-voltage Relationship

The difference between the two curves is a function of the junction temperature. The bottom curve, 25°C, describes a nearly perfect logarithmic function from near zero diode current to almost 1 mA. At 75°C, the top curve, the logarithmic function starts at about 1 nA. For a given value of diode current, the voltage across a silicon diode decreases about 2.4 V/°C rise in temperature. This is, of course, a limitation on using a diode for generating logarithmic functions. In highly critical circuits, we often find such diodes encased in a cooler to maintain them at a constant temperature. Generally speaking, the lower the temperature, the greater the logarithmic range of the diode. In less critical applications, we may find various types of temperature compensation circuits instead of a cooler.

If we superimpose a true logarithmic line over each of the curves, we find that these curves are only approximations. They deviate from the true logarithmic line on both ends, and to a lesser degree, along the center. The deviations on the low end are caused by the fact that current and voltage must reach zero at the same time. The deviations on the high end are due to the junction resistance of the diode. Slight deviations along the center of the curve are explained by the variations in factor m as the current changes.

Zener Diode The Zener diode exhibits nonlinear reverse characteristics, over a small operational area, that are nearly logarithmic. The Zener voltage approximates a logarithm of the current when the Zener is operated in the transition region between the forward region and the Zener region. This region is labeled A on the diagram in Fig. 9–2.

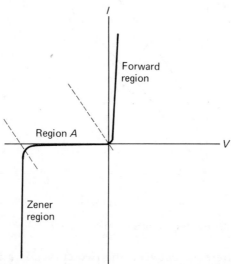

FIGURE 9–2. Zener Current-voltage Relationship

Transistors Nearly all types of transistors exhibit nonlinear characteristics which can be made useful. The silicon bipolar transistor exhibits a logarithmic relationship similar to the junction diode when we hold the collector voltage at a constant value. This characteristic can be enhanced, and spread over a wider range of current, by using a feedback circuit.

Both triode and tetrode transistors have nonlinear ac transfer characteristics when we allow the input signal to control either the direct voltage or current. At low frequencies, the transistor parameter h_i varies inversely with the dc, and gain can be made to vary inversely with h_i. To state the effect in a more direct fashion, gain is directly proportional to direct current. This characteristic is useful in amplitude compression and expansion circuits. At high frequencies, the parameter h_f varies with the collector direct current. This variation in h_f can be used in automatic gain control circuits to make the gain a function of the input signal amplitude.

A silicon unijunction transistor shows a decrease in interbase resistance as the emitter current increases. This variation is caused by conductivity modulation of the silicon in the interbase region. This characteristic is very useful in amplitude compression and expansion circuits, where the unijunction transistor is used as a variable resistance. The interbase resistance is less than 100 ohms with 10 mA of current. This resistance rises to about 8000 ohms as the current drops to zero. The interbase resistance is linearly related to the emitter current at currents between 1 and 5 mA.

When we use the unijunction transistor in high frequency circuits (higher than 1 MHz), the interbase resistance is independent of the base-to-base voltage. This enables the construction of automatic gain control for very critical pulse circuits.

The field effect transistor becomes a voltage-variable resistor at voltages below pinch-off. When it is operated in this region, we can obtain a change in the dynamic resistance from drain to source by making a corresponding change of the gate bias. This characteristic is very effective in automatic gain control circuits.

The field effect transistor is also a natural amplitude multiplier when used in circuits for pulses of very short pulse duration. Multiplication is based on the relationship of the channel conductance to the base voltage. The simplest way to accomplish this is to use a bridge circuit composed of two field effect transistors. Two variables are used to control conductance in such a way that the difference in channel currents is proportional to the product of the two variables.

Varistors We call a voltage-sensitive resistor a varistor, and it has useful nonlinear characteristics. A varistor is formed by pressing together a mixture of silicon carbide granules and a suitable binding agent. The mixture is then heated to a high temperature.

222 Nonlinear Functions

The following equation expresses the nonlinear qualities of such a varistor:

$$I = kV^n$$

where I = instantaneous current, K is a constant derived from the physical qualities of the device, V = instantaneous voltage applied, and n is an exponent between 1 and 5 according to the value of k. (As k goes up, n goes down.) These nonlinear characteristics can be put to use as controls in circuits that are designed to expand or compress the signal amplitude.

We can use the input voltage to control the resistance of the varistor, and let the varistor current control the gain of an amplifier. When used in this fashion, the output voltage of the transistor is approximately equal to the cube root of the input voltage.

Thermistors A thermistor is a temperature-controlled resistor. It generally is constructed by blending the oxides of manganese, nickel, and cobalt. The particular blend determines the specific resistance and negative temperature coefficient of the thermistor. Thermistors are frequently used as control devices in attenuation circuits and negative feedback circuits. The temperature can be controlled by the signal amplitude. As signal amplitude goes up, temperature goes up, and resistance goes down. Of course, the thermistor can be constructed so that it has a positive temperature coefficient. The thermistor current can be used to control the gain of an amplifier. Temperature-compensated square law circuits can be constructed by using combinations of thermistors and varistors.

Vacuum Thermocouples The vacuum thermocouple is another element with useful nonlinear characteristics. These devices produce an output voltage proportional to the heating effect of the applied input signal. This effect gives us a square law relationship between the input and the output. The vacuum thermocouple may be obtained to fit nearly any type of circuit. These devices are useful from dc to a frequency in excess of 300 MHz, and they have a range of heater currents from 3 mA to 1 A. The wide frequency response of vacuum thermocouples make them ideally suited for wide band applications.

Optoelectronic Devices When we combine a photoelectric cell with a controlled light source, we have an optoelectronic device. This combination is mounted in a light-tight case, and they provide ideal gain control characteristics for many applications. These devices have a cell resistance that can be controlled over a wide range of values. We can use the device as a control over a frequency range from dc to several hundred kHz. The ionized gas light source can be used to control the response time; it can be made as short as a few milliseconds.

Servomechanisms Servomechanisms provide us with a major method of achieving nonlinear transfer characteristics. We can use these devices to establish nonlinear relationships between input and output, and we frequently find them used as amplitude multipliers. The single term *servomechanism* includes potentiometers, servo amplifiers, servo motors, gears, and limit stops. The potentiometers are primarily responsible for the nonlinear characteristics. The potentiometers usually have taps welded to the turns of wire or else have the wire wound so that we have a nonlinear relationship between shaft rotation and resistance. Precision mechanical components are required for accurate nonlinear amplifier systems.

FUNCTION GENERATORS

Control circuits may be constructed to provide various types of nonlinear functions. We call these circuits nonlinear function generators. These nonlinear functions that can be generated range from logarithmic to square law. Some of these circuits are used as simple generators, and others are used in combination with linear amplifiers.

Logarithmic There are several types of circuits that we can use to give us an output signal that is proportional to the logarithm of the input signal. We call these circuits logarithmic function generators. The logarithmic attenuator is an example of this type of circuit. Most logarithmic function generators may be used to obtain an antilog function, and we frequently find them in analog multiplication and division circuits. Figure 9–3 is a schematic for a logarithmic attenuator.

Branches 1 through 6 are composed of a dropping resistor and a silicon crystal diode. The resistance values are chosen so that each resistance is 10 times the value of the resistor on its right. For instance: R1 may be 150 MΩ. In this case, R2 is 15 MΩ, R3 is 1.5 MΩ, R4 is 150 kΩ, R5 is 15 kΩ, and R6 is 1.5 kΩ.

In a static condition, each of the diodes is reverse-biased.

224 Nonlinear Functions

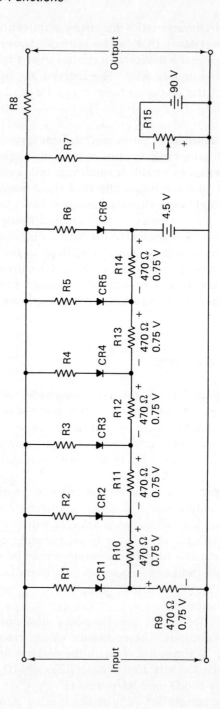

FIGURE 9-3. Logarithmic Attenuator

Our bias is supplied by a 4.5-V battery and six 470-ohm resistors. The reverse bias on CR6 is 4.5 V. This bias value decreases by 0.75 V with each diode until CR1, which has a bias of 0.75 V. When our input current is at a low level, all six branches are closed. R7 is about 10 times the value of R8 so that the small values of current flow through R8 to a low impedance in the following stage.

As our input current increases, the potential at the top of each branch becomes more positive. When this potential reaches 0.75 V, CR1 starts to conduct. When it reaches 1.5 V, CR2 starts to conduct. This continues until a potential of 4.5 V is reached. At this time, all of our diodes are forward-biased and conducting. Some of the input is diverted through each of the conducting paths. In each branch the amount of diverted current is determined by the resistance of the conducting branch with respect to the resistance of R8.

The 90-V battery in conjunction with R15 provides an adjustment to compensate for diode leakage current. Without this compensation, leakage current through the six diodes would cause an output current through R8 when we have no input. R15 is adjusted (with zero input) until we have a zero output current.

The input-output current characteristics of this attenuator describes a nearly perfect logarithmic relationship. We can come even closer to the true logarithmic characteristic by adding additional parallel branches.

A logarithmic function generator of moderate accuracy is shown in Fig. 9-4.

In a static condition, all the diodes are forward-biased and are, in effect, shorted out. With low values of input voltage, the equivalent resistance of R1 through R7 is the value of the input loading resistance to the amplifier. In this condition, the signal has a minimum attenuation and the amplifier is operating at full gain. As our input signal voltage increases in a positive direction, each diode, in turn, becomes reverse-biased and opens its particular branch. Each time a branch opens, the equivalent resistance increases with a corresponding decrease in gain. When all branches are open, the loading resistance is the value of R1. In this condition, we have maximum attenuation and the amplifier is operating at minimum gain. The seven changes in input loading resistance provides a seven-step approximation to a logarithmic curve of the input to output voltage characteristic. With input voltages of 1 to 10 volts and output voltages from 0 to 10 volts, this logarithmic function generator has an inherent error of about ±0.5 percent.

226 Nonlinear Functions

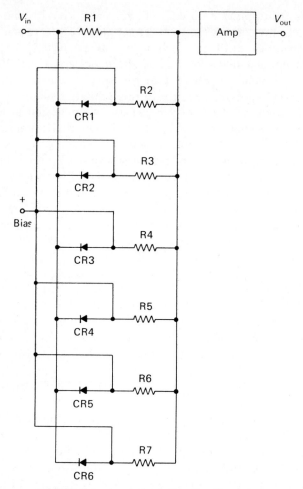

FIGURE 9-4. Logarithmic Function Generator

Square Law A slight variation of the logarithmic function generator circuit produces a circuit for a square law function generator, as shown in Fig. 9-5.

Now we have the reverse situation; with low input voltage, all diodes are cut off (open) because of reverse bias. The input loading resistance is at its maximum value (the value of R1), and the amplifier is operating at minimum gain. As our input voltage increases in amplitude, the diodes become sequentially forward-biased and add resistors in parallel with R1. As each branch is added, our input loading resistance becomes smaller, and the amplification becomes greater. When our input is at its maximum amplitude, all

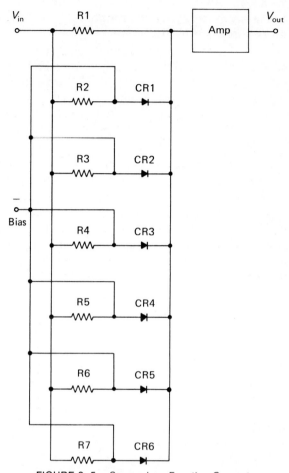

FIGURE 9-5. Square Law Function Generator

the diodes are conducting, the input loading resistance is at a minimum value, and the amplifier is operating at maximum gain. The seven-step change in our input loading resistance gives us a seven-step approximation to a square law curve of the output to input voltage characteristic.

Servo Techniques Several nonlinear functions can be generated by use of servo techniques. The servo system may be either dc or ac. The dc system consists of drift stabilizing amplifier input to a servo amplifier output stage. The output stage drives a dc motor. The ac system is driven by a two-phase ac motor. One winding is controlled by a fixed current, and the other is excited by

228 Nonlinear Functions

the output of a servo amplifier. A simplified block diagram of a servo system is illustrated in Fig. 9–6.

Batteries V1 and V2 are dc source voltages for input and output, respectively. The setting of potentiometer R1 determines the level of dc input to the servo amplifier. The amplitude of this input indirectly determines the number of degrees that our servo motor will rotate. As the motor turns, it alters the setting of potentiometer R4. So the angular position of the motor indirectly determines the potential between the center arm of R4 and ground, and this potential is the output voltage. Notice that in all cases, the voltage between the arm of R4 and ground is the same voltage that appears across R5, and R5 is our output load resistor. The amplitude of V_{out} is approximately equal to V_{in} (R3/R2), and in no case can it ever exceed the value of $V2$. Nonlinear characteristics in this system can be achieved by nonlinear resistors for either or both R2 and R3. We could also accomplish nonlinearity by allowing the servo motor to vary the resistance of either or both R2 and R3.

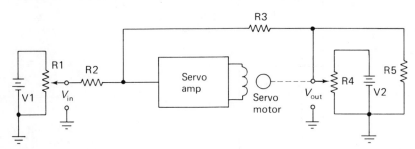

FIGURE 9–6. Simplified Servo System

As previously stated, the potentiometers used in a servo system are the primary sources of nonlinearity. Simple relationships can be approximated by adding a load (pad) resistor to an untapped linear potentiometer. But if we desire accurate logarithmic or square law functions, we must use either tapped linear potentiometers or nonlinear resistive elements. Figure 9–7 illustrates two methods of tapping and padding a linear potentiometer.

In Fig. 9–7(a), the variable padding resistors are connected in series. In section (b), each of the padding resistors is in parallel with several sections of the potentiometer, and to some extent, with one another. The arrangement in (a) is safer from the standpoint of possible burnout of the potentiometer, and it draws minimum current from the power source. The arrangement in (b) is more versatile in the type of functions that it can produce. However, this arrangement draws more current from the power source, and each tap must be fused to avoid burnout of the potentiometer.

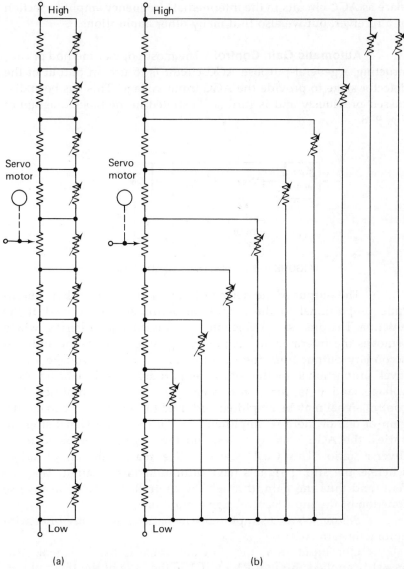

(a)   (b)
FIGURE 9-7. Tapped and Padded Potentiometers

LOGARITHMIC AMPLIFIERS

A common application of the logarithmic function is the automatic gain control circuit. Logarithmic amplifiers can be designed to compress or expand the gain in accordance with the amplitude of an input signal. This compression and/or expansion generally takes place as AGC circuits in the intermediate frequency amplifier section of a receiver, but we also find many other applications.

Automatic Gain Control The most popular method of constructing a general-purpose AGC circuit is to use an output of the detector stage to provide the AGC input voltage. This has been discussed previously and is further illustrated in the block diagram of Fig. 9-8.

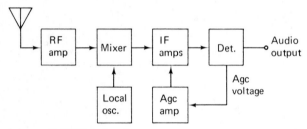

FIGURE 9-8. Obtaining Input AGC Voltage

The output of the detector is an audio signal with an amplitude proportional to the amplitude of the signal received at the antenna. The process of detection is rectification and filtering, which removes the intermediate frequency signal to produce the audio. A secondary output from the detector can be a direct voltage with a level proportional to the signal amplitude. This is labeled AGC voltage, and it is our control signal for the AGC amplifier. This control signal may be amplified and used to compress or expand the gain of one or more IF amplifiers. As our incoming signal strength varies, the AGC voltage varies, and the IF gain is adjusted in an inverse ratio. The weak signals have maximum amplification, whereas the strong signals have minimum amplification. The current, and thus the gain, through the controlled stage is an inverse function of the input control signal amplitude.

Figure 9-9 is a simplified schematic of an IF amplifier with automatic gain control.

Our input from the previous stage is the IF signal. This signal is applied through R1 and C1 to the base of the IF amplifier, Q1. The output of Q1, still an IF signal, is developed across R3 and coupled to the next stage through C2. The detector processes this IF signal and couples back an AGC voltage proportional to the signal amplitude. The signal current divides between the ac resistance of diode CR1 and the ac input resistance of Q1.

In this case, we are using a positive AGC voltage, which is applied across R2 as forward bias for CR1. The conductance of CR1

Logarithmic Amplifiers 231

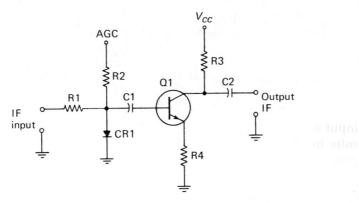

FIGURE 9-9. AGC Amplifier with Diode Shunt

is directly proportional to the AGC level, and its ac resistance is inversely proportional to this same level. As our input signal strength increases, the AGC voltage becomes more positive and causes CR1 to conduct more. The increased conduction of CR1 reduces its ac resistance and allows more of the signal current to pass through the diode. As more signal current passes through the diode, this leaves less signal current through the junction of Q1, which lowers the gain of the stage.

We must keep in mind that this AGC action is not an instantaneous thing. A finite period of time is required between a change in signal amplitude and a corrective AGC action on the amplifier. In some cases, we have additional delays deliberately built in to slow down the response time.

We may also achieve amplitude expansion or compression by using a transistor as a variable resistor in conjunction with the AGC voltage. This technique is often used with low and medium frequencies. The AGC voltage serves as a control signal to vary the dynamic resistance of a bipolar transistor. This technique is illustrated in Fig. 9-10.

The principles here are much the same as in our previous circuit. The primary difference is the use of a transistor instead of a diode. Transistor Q1 is a variable shunt resistor whose ac resistance is inversely proportional to its conductance. The conductance is controlled by the positive AGC voltage on the base of the transistor. The conductance is directly proportional to the AGC voltage level.

Our input signal is divided between the dynamic resistance of Q1 and the ac input resistance of Q2. C1 completes the signal path between the base of Q2 and the collector of Q1, and at the same time, blocks the direct voltage from the base of Q2. An increase in

232 Nonlinear Functions

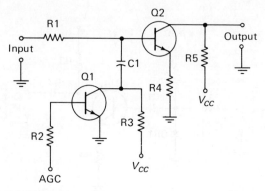

FIGURE 9-10. AGC Amplifier with Transistor Shunt

input amplitude causes an increase in output amplitude, which results in a higher level of AGC voltage. The higher level of AGC voltage shunts more of the signal and reduces the gain of the stage. Thus gain is inversely proportional to the amplitude of the input signal.

DC Amplification We often encounter logarithmic dc amplifiers in electronic instrument circuits. The amplifiers are used to compress a positive input current into a proportionate direct voltage. These circuits frequently combine the logarithmic characteristics of junction diodes with the variable input impedance of field effect transistors. A sample circuit is shown in Fig. 9-11.

The input voltage is applied across CR1 and the high input resistance of field effect transistor Q1. The leakage current of CR1 is negligible over the operating range, which results in a linear amplification of the diode voltage. The output voltage then becomes a logarithmic function of the input current.

Transistor Q2 is a variable source resistor for Q1. The feedback through R1 and R2 varies the resistance of this transistor which controls the operating source current for Q1.

Transistors Q3 and Q4 compose a difference amplifier that acts only to a change in dc. R7 and R8 control the base potential on Q4, and R8 is adjusted for a quiescent voltage output (from Q5) in the vicinity of zero. These transistors also compensate for voltage drift during current changes.

Constant Output AC Amplifier Amplifiers with a constant output amplitude regardless of input level are useful in AGC circuits and in many other applications. Such an amplifier is illustrated in Fig. 9-12.

Logarithmic Amplifiers 233

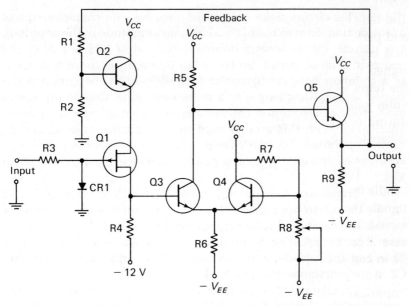

FIGURE 9-11. Logarithmic DC Amplifier

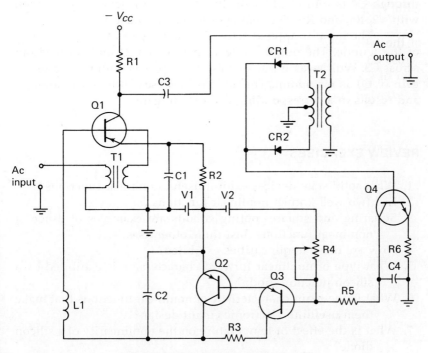

FIGURE 9-12. Constant Output Amplifier

234 Nonlinear Functions

This circuit makes use of the variable gain characteristics of a tetrode transistor to maintain a constant amplitude in the ac output. The tetrode, Q1, is serving double duty; part of it is in a dc circuit and part in an ac circuit. So far as our dc circuit is concerned, Q1 is in a common base configuration. The dc part of the circuit is designed to keep dc changes to a minimum value. Our ac input sees the tetrode as a common emitter amplifier that is capable of maximum gain. With this arrangement, we can change both h_{fe} and h_{fb} during operation, but the change ratios are considerably different. Whereas h_{fb} changes by a ratio of 4:3, h_{fe} changes by a ratio of about 15:1.

Transistors Q2 and Q3 form a dc difference amplifier to handle the difference signal between the ac output and the reference signal. The combined collector currents of Q2 and Q3 provide transverse bias for Q1. The dc amplifier must be isolated from the ac on base 2 of the tetrode, and this function is performed by capacitor C2 in conjunction with inductor L1.

Transistor Q4, with its associated circuits, provides control amplification. Diodes CR1 and CR2 are rectifiers that provide full-wave rectification of the ac on the secondary of T2. The current through Q4 is controlled by the dc from the rectifiers in conjunction with V2, R4, and R5. The reference is set by adjusting R4.

The input is applied across T1 and provides ac bias for base 2 of the tetrode. The output is developed across R1 and coupled out across C3. With weak input signals, the ac bias is minimum, and the gain of Q1 is maximum. The ac bias increases with signal strength and results in a corresponding decrease in gain.

REVIEW EXERCISES

1. What solid state devices exhibit useful nonlinear functions?
2. List two well-known nonlinear functions.
3. Squaring and square rooting circuits are examples of usage of nonlinear functions. List three other uses.
4. How are these nonlinearities exhibited?
5. What type of nonlinear forward characteristic is exhibited by a silicon junction diode?
6. What is the main characteristic of nonlinear functions that make them useful in electronic circuit design?
7. What is the effect of temperature on the nonlinearity of a silicon diode?

8. List two ways of overcoming the effect of temperature on a diode.
9. Describe the nonlinear relationship of reverse voltage and current for a Zener diode.
10. Draw a voltage-current curve for a Zener diode and label the region where the nonlinear characteristics appear.
11. What can be done to a silicon bipolar transistor to bring out its logarithmic characteristics?
12. What can be done to improve and extend the range of the characteristic in item 11?
13. When do triodes and tetrodes exhibit nonlinear ac transfer characteristics?
14. In triodes and tetrodes, what is the relationship of dc, h_i, and gain at low frequencies?
15. In triodes and tetrodes, what is the relationship of h_f and dc at high frequencies?
16. What type of nonlinearity is produced in a silicon unijunction transistor by increasing the emitter current?
17. How do we account for the variation of item 16?
18. What is the approximate interbase resistance of a silicon unijunction transistor with an emitter current of
    a. 0?
    b. 10 mA?
19. At what range of emitter currents do we have a linear relation between current and interbase resistance in a silicon unijunction transistor?
20. Under what conditions does a field effect transistor become a voltage variable resistor?
21. Under what conditions does the field effect transistor become an amplitude multiplier?
22. What is a varistor?
23. Describe how the varistor may be used in amplitude expansion and/or compression circuits.
24. What is a thermistor?
25. Describe how the thermistor can be used in amplitude expansion and/or compression circuits.
26. What is the relationship between the output voltage and the input signal of a vacuum thermocouple?
27. What is the useful range of vacuum thermocouples in terms of
    a. Frequency?
    b. Heater current?
28. What is the purpose of a logarithmic generator?

236 Nonlinear Functions

29. When one is using a diode logarithmic generator to control the gain of an amplifier, what is the relationship of input voltage, input loading resistance, and amplifier gain?
30. When one is using a diode square law generator to control the gain of an amplifier, what is the relationship of input voltage, loading resistance, and gain?
31. What components are the primary sources of nonlinearity in a servo system?
32. What is the most common source of the control signal for an IF AGC circuit?
33. Describe the relationship of signal amplitude, AGC voltage level, and controlled amplifier gain in an AGC circuit.

Chapter Ten

Sinusoidal Oscillators

Electronic equipment is useless machinery without the proper operating power and the proper input information. A major portion of electronics deals with devices that are solely designed to satisfy these two requirements. Our electronic equipment requires alternating currents with a frequency range from a few hertz to several THz (THz = 10^{12}). The low frequencies can be supplied by ordinary electric generators with rotating armatures. These generators can supply currents at frequencies up to a few thousand hertz.

When our frequency needs exceed the practical limits of mechanical rotating devices, we must turn to an electronic generator. Electronic generators have been used to produce usable currents at frequencies in the range of visible light in excess of 1000 THz. We call these electronic generators oscillators. An oscillator is an electronic circuit that converts direct current to alternating current. The frequency of oscillation is determined by the values of the components that we use in the circuit.

We have a great variety of oscillators. Our discussion must be limited to several types that you are most likely to encounter in ordinary electronic environments. In this chapter we deal with the principles of oscillation and transistor oscillator circuits that produce ac in a sinusoidal waveform.

PRINCIPLES OF OSCILLATION

Our oscillator circuit must contain four essential sections in order to produce sustained oscillations. These are

1. A dc power supply.
2. An amplifier.

3. A frequency determining device.
4. Positive (regenerative) feedback.

Figure 10-1 shows the relationship of these sections.

The dc power source that supplies bias for the transistor amplifier is the same as that required for any other amplifier. This has been adequately covered in previous chapters. The only difference here is that we must have a steady dc source. This is often accomplished by taking bias voltages from a regulated power supply. We shall examine the other three requirements in some detail in the following paragraphs.

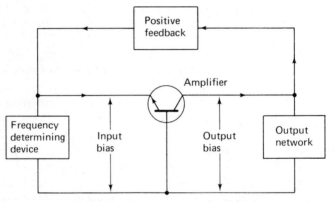

FIGURE 10-1. Requirements for Oscillation

The Amplifier A transistor can be used as our amplifying element for an oscillator at frequencies between cutoff frequency and maximum frequency (f_{max}) of the transistor. The cutoff frequency is the frequency that produces a current gain of 0.707 times its low-frequency value. The low-frequency value is normally measured at 1000 Hz. The maximum frequency of oscillation is designated as f_{max}, and it is the frequency where the transistor power gain is reduced to unity when the transistor is used in a common emitter configuration. With a power gain of unity, we have the same power in the output as we have in the input. Obviously, we shall have some losses in our feedback circuit, so with a power gain of unity, our transistor cannot sustain oscillations.

We may construct oscillator circuits by placing a transistor in any of the three basic amplifier configurations, and all three are used. The particular configuration chosen for a particular oscillator is determined by the circuit requirements. Let us consider the three configurations from the standpoint of oscillator amplifiers.

The common base configuration has the lowest input impedance and the highest output impedance. This difference in impedances makes it difficult to construct a feedback circuit without excessive energy losses. If we cannot match impedances, we must compensate for the energy losses in the feedback circuit. The common base has a reasonable gain of both voltage and power, but the current gain is less than unity. There is no phase reversal between input and output.

The common emitter configuration has moderate input and output impedances. This reduces the requirements for matching impedances in the feedback circuit. The current and voltage gains are reasonably high, and the power gain exceeds that of both the other configurations. There is a phase shift of 180 degrees between the input and the output.

The common collector configuration has a high input impedance and a moderate output impedance. We have the same impedance matching problem here that we have with the common base, if we are to have feedback without excessive losses. The current and power gains are reasonably high with this configuration, but the voltage gain is less than unity. We have no phase reversal between the input and the output.

Frequency Determining Device Many of our sinusoidal oscillator circuits are based on the principle of an electric tank circuit, which we call the flywheel effect. This effect describes the interchange of energy between the electrostatic field of a capacitor and the electromagnetic field of an inductor. Consider the circuit of Fig. 10-2.

In part (a) of the figure, current I from the battery is charging the capacitor with the polarity indicated. After our capacitor is fully charged, we open the switch, as indicated in part (b). Our capacitor now takes the role of a power supply and causes current through the inductor, as indicated by the arrows. As the capacitor loses its charge, the inductor is charging as indicated as a result of the expanding magnetic field. When the charge equalizes between the plates of the capacitor, current would cease except for the inductor action. However, when capacitor current tries to stop, the magnetic field around the coil starts to collapse. This collapsing field reverses the polarity on the inductor and keeps the current going in the same direction until the magnetic field is fully discharged. This inductor action charges the capacitor, as indicated in part (c). Now when the magnetic field is fully decayed, the capacitor again takes over and forces current back in the opposite direction.

240 Sinusoidal Oscillators

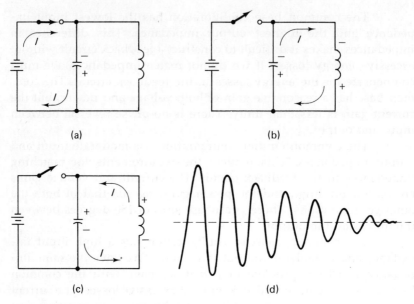

FIGURE 10-2. Flywheel Effect

Part (d) illustrates the waveshape of the current in the tank circuit. The capacitor discharges on the first quarter-cycle and the coil charges during this time. The coil discharges during the second quarter-cycle and charges the capacitor in the opposite direction. The capacitor reverses the current, discharges again and recharges the inductor during the third quarter-cycle. The coil takes over again for the fourth quarter-cycle and recharges the capacitor in the original direction. The action is repeated for each succeeding cycle. At the end of each cycle, the capacitor is charged slightly less than it was on the last cycle. This decreasing charge is caused from losses in our circuit, and it results in the gradual decay in the amplitude of the current waveshape.

If we could close the switch at the end of each cycle just long enough to replace the circuit losses, the tank circuit would produce continuous sine waves of current at a constant amplitude. We cannot do this with a mechanical switch, but our feedback circuit can accomplish the same effect in an oscillator circuit.

The frequency of the sine wave from the tank circuit is the natural frequency required for a complete reversal of the charge in this tank circuit. This frequency is the resonant frequency of the tank and is dependent upon the values of the capacitor and inductor. The resonant frequency is

$$f_r = \frac{1}{2\pi \sqrt{LC}}$$

Feedback Our feedback signal must be strong enough to replace all the energy losses in the circuit in order to sustain oscillations. The output power from the amplifier is divided. A portion enters the feedback circuit and the remainder is usable energy to be coupled into another stage. To avoid extreme energy losses in the feedback network, the input to the network must match the output impedance of the amplifier, and the terminal end of the network must match the input impedance of the amplifier. If we have our amplifier in a common emitter configuration, our feedback network must provide a phase shift of 180 degrees to compensate for the phase shift through the transistor. If the amplifier is in either a common base or a common collector, the feedback network must have no phase shift, because the input and output are already in phase.

Frequency Stability In most cases, a steady fixed frequency of oscillation is a basic oscillator requirement. Many factors affect the frequency. One of these is the dc operating point; it must be fixed on a linear portion of the transistor's characteristic. Nonlinear operation will result in parameter changes and frequency shifts. This is why the dc bias must be from a steady source. Temperature changes tend to shift the operating point, so temperature stabilization is essential.

The collector to emitter capacitance is another frequency problem in our oscillators. This capacitance changes with voltage changes or temperature variations. The capacitance is more troublesome as the frequency increases. At high frequencies, we must use an external capacitor across the collector-emitter electrodes. Adding this capacitor in such a fashion places it in parallel with the internal capacitance; this increases the total capacitance, and reduces the capacitive reactance.

RESONANT FEEDBACK OSCILLATORS

Oscillators may be either self-excited or triggered, and they may have either resonant or nonresonant feedback circuits. We are considering only self-excited oscillators in this chapter, and in this section, we shall analyze some common oscillator circuits that use resonant feedback.

Basic Oscillator First we shall examine a greatly simplified oscillator circuit. In this explanation, we shall omit the exact biasing and stabilizing arrangements for simplicity. We want to illustrate

242 Sinusoidal Oscillators

the basic component arrangement and polarity applications in transistor oscillators. A very basic oscillator circuit is shown in Fig. 10–3.

Here we have a transistor in a common base configuration as our amplifier. The collector load is composed of an LC tank circuit (C1 and L2) that is placed across the output. This tank circuit is our frequency determining device. L1 is across the input to our transistor and provides the feedback to keep the oscillations going. L1 and L2 are generally two windings on a single transformer. The frequency is determined by the values of C1 and L2. We quite often refer to coils such as L1 as tickler coils.

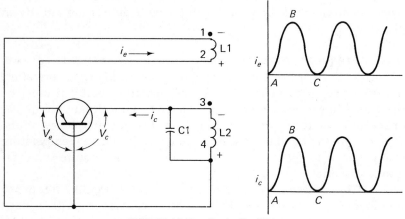

FIGURE 10–3. Basic Oscillator

Point A on the current waveforms represents the point where power is applied and the start of oscillation. As power is applied, our transistor is forward-biased and starts to conduct. Collector current from pins 3 to 4 through L2 produces the polarity indicated on both L2 and L1. Between points A and B, both collector current and emitter current increase at a sinusoidal rate. This increase is aided by the negative potential from L1 to the base of the transistor which grows stronger as current increases. This is positive (regenerative) feedback, because it aids the action of the transistor.

At point B, our transistor is saturated and both collector and emitter currents cease to increase. At this time feedback ceases because of a steady current through L2. Loss of feedback causes the emitter current to begin to decrease which decreases the collector current. The potential on L2 reverses to oppose the decrease in collector current, and this switches the polarity on the feedback coil. The positive potential from L1 that is now applied to the base keeps the emitter current decreasing at a sinusoidal rate from point B to point C. The collector current follows this same change. This positive from L1 is still regenerative feedback; emitter current was decreasing, and this potential aids that decrease.

At point C, all current ceases because our transistor is cut off by reverse bias. When i_c ceases, feedback ceases, and without feedback, our transistor becomes forward biased and starts conducting again.

Our transistor is alternately driven from cutoff to saturation. The resonant frequency of the tank circuit determines the frequency of operation and thus controls the time for each complete cycle of events.

Tuned Base Armstrong Oscillator One of our simplest transistor oscillators is the tuned base *Armstrong* oscillator which we sometimes call a *tickler coil* oscillator. A schematic for this oscillator is illustrated in Fig. 10–4.

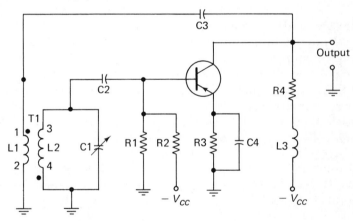

FIGURE 10–4. Tuned Base Armstrong Oscillator

In this application we have a transistor in a common emitter configuration, and we use a single source of dc for both input and output bias. R1 and R2 form a voltage divider to give us proper input bias. R4 is our collector load resistor. L3 is a radio frequency choke (RFC) coil to prevent the oscillations from entering the dc power supply. C3 is a dc blocking capacitor to isolate the dc from the oscillating components; C2 is another blocking capacitor that serves the same function. R3 is a swamping resistor for stability, and it is bypassed by C4 to prevent signal degeneration. The frequency determining device is a tuned tank circuit between base and ground composed of L2 and C1. L1 is the tickler coil that feeds a portion of the output back into the tank circuit. The frequency is determined by the values of L2 and C1. C1 is a variable capacitor that can be used to vary the frequency. Sometimes we find the capacitor fixed and a variable slug in the inductor for performing the frequency adjustments.

When we apply power, our transistor is forward-biased and current starts. At this instant the full $-V_{CC}$ is felt at the collector. In the first half-cycle, i_c rises from cutoff to saturation. The collector potential becomes less negative as the increasing current drops more voltage. At saturation (end of first half-cycle) the collector voltage is zero.

The positive going change at the collector is coupled across C3 to L1. The positive at L1 pin 1 is inverted across the transformer and applied to the base of our transistor as an increasing negative potential. This is positive feedback; it aids the forward bias on the transistor and drives it into saturation.

At the instant of saturation feedback ceases, forward bias decreases, and collector current starts to decrease. The collector potential now starts to change in a negative direction. This negative change is coupled to L1 pin 1 as a negative and to the transistor base as a positive. Our feedback is now opposing the bias and aiding the decreasing current. During this second half-cycle, the current decreases from saturation to zero. The second half-cycle terminates when the transistor becomes reverse-biased and all current ceases. When collector current ceases, feedback ceases. At that instant, reverse bias is removed and the transistor starts conducting again. The oscillations will continue indefinitely so long as the feedback replaces the energy losses in the circuit. This assumes, of course, that other factors remain normal.

Tuned Collector Armstrong Oscillator We find that the tuned collector Armstrong oscillator is a much favored oscillator circuit. One reason for this popularity is the ease of solving the impedance matching between output and input. This circuit is schematically illustrated in Fig. 10–5.

Now we have our frequency determining tank circuit in the collector, and it also serves as the collector load impedance. The frequency is determined by the values of C3 and L2, and is adjustable by varying C3. L1 is our feedback coil, and the impedance matching problem is solved by selecting a proper turns ratio between L2 and L1. The transformer coupling from L2 to L1 accomplishes a 180-degree phase shift between collector and base to compensate for

Resonant Feedback Oscillators 245

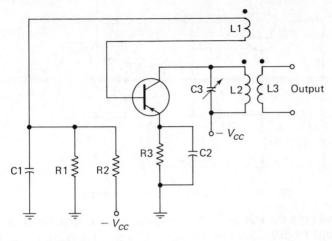

FIGURE 10-5. Tuned Collector Armstrong Oscillator

the phase shift through the transistor. C1 provides an ac ground for one side of the tickler coil.

The forward bias for our transistor is provided by the voltage divider consisting of R1 and R2. Notice that the potential on the junction of R1 and R2 is applied through L1 to the base of the transistor. The transistor operation is stabilized by the swamping resistor, R3. R3 is bypassed by C2 to prevent degeneration of the signal.

The output signal is a sine wave that couples from L2 to L3. The transistor is alternately driven from cutoff to saturation at the resonant frequency of the tank circuit.

Common Base Colpitts Oscillator The common base Colpitts oscillator uses a tank circuit that is fed from the collector without being the collector load impedance. Our feedback, in this case, is taken from the tank circuit back to the emitter. Figure 10-6 is a schematic of this oscillator.

Our feedback is now accomplished by tapping the capacitor side of the output tank circuit. The energy tapped between capacitors C2 and C3 is returned to the input as regenerative feedback. During one half-cycle, the polarities on C2 and C3 will be as indicated in the illustration. Since the voltage across C3 is the feedback potential, this gives us a positive on the emitter to aid the increasing current. On alternate half-cycles, these polarities will reverse, and the negative voltage across C3 will then aid the decreasing current. Just as the tickler coil is a mark of identification for the Armstrong, this tapped capacitor feedback identifies the Colpitts oscillator.

Either or both capacitors in the tank circuit may be adjusted

246 Sinusoidal Oscillators

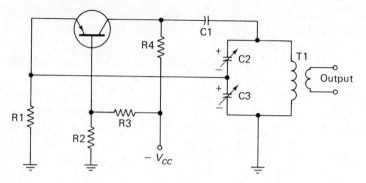

FIGURE 10-6. Common Base Colpitts Oscillator

to control both the frequency and the level of feedback. For minimum loss in our feedback circuit, the ratio of capacitive reactance of C2 to C3 should approximately equal the ratio of the output impedance to the input impedance of the transistor.

Common Emitter Colpitts Oscillator The common emitter configuration is used more often in the Colpitts oscillator than the common base. It involves only a slight rearranging of the circuit components. The Colpitts feedback from tapped capacitors is still readily apparent. This circuit is illustrated in Fig. 10-7.

Our frequency determining tank circuit consists of the combination C1 and C2 and the primary of T1. The output is taken across the secondary of T1. The tank circuit is connected between the collector and the base. C3 and C4 are ac coupling capacitors which also

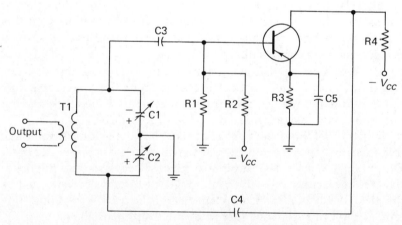

FIGURE 10-7. Common Emitter Colpitts Oscillator

block the dc from the oscillating circuits. R1 and R2 form a biasing voltage divider. R3 is our usual stabilizing swamping resistor, and C5 bypasses this resistor to prevent signal degeneration.

On one half-cycle, the collector current is increasing and moving the collector potential in a positive direction. During this alternation the indicated polarities are developed on C1 and C2. The potential across C1 is the actual feedback across the input junction, and it is 180 degrees out of phase with the collector voltage. During alternate half-cycles, the polarity of feedback is reversed to place a positive potential between base and emitter. This positive potential aids the decreasing current.

Shunt Fed Hartley Oscillator We have a marked similarity between the Colpitts and the Hartley oscillators. The primary difference is the means of tapping the feedback from the tank circuit. The Colpitts oscillators use tapped capacitors for obtaining the feedback. The Hartley oscillators use a tapped inductor for this purpose. A Hartley oscillator is illustrated in Fig. 10-8.

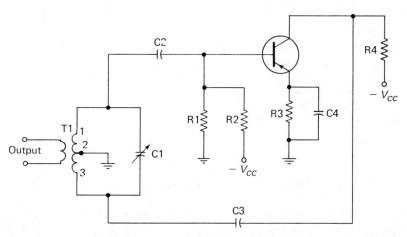

FIGURE 10-8. Shunt Fed Hartley Oscillator

Notice how closely this circuit resembles that in Fig. 10-7. The only difference is a tapped inductor instead of tapped capacitors. The frequency is determined by C1 and the total primary winding of T1. We still use a variable capacitor to adjust the frequency, but we use a portion of the voltage across the inductor as feedback. The inductor voltage between T1 pins 1 and 2 is the potential between base and emitter; this is our feedback voltage.

Pin 1 of T1 is coupled to the base of our transistor, while

pin 3 is coupled to the collector. Since the potentials at these two pins are always 180 degrees out of phase, this compensates for the 180-degree phase shift through our transistor.

During one alternation, pin 3 of the transformer goes in a positive direction while pin 1 goes in a negative direction. The increasing negative on the base aids the increasing current. On alternate half-cycles, pin 3 is negative and pin 1 is positive. The increasing positive on the base aids the decreasing current. Notice that the dc is completely isolated from the oscillating circuits by C2 and C3, which couple the ac signals.

Series Fed Hartley Oscillator With a very slight modification of the circuit, we can convert from the shunt fed Hartley oscillator to a series fed Hartley oscillator. This oscillator circuit is illustrated in Fig. 10–9.

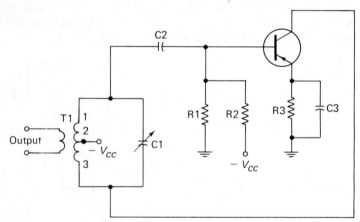

FIGURE 10–9. Series Fed Hartley Oscillator

If you compare this circuit to that in Fig. 10–8, you should notice that we have left out the collector load resistor, eliminated the coupling capacitor between collector and tank, and slightly rearranged the collector voltage. Now our collector load impedance is the portion of T1 primary between pins 2 and 3. Our dc voltage for bias is now mixed with the ac oscillations in the tank circuit. This undesirable feature is the chief objection to the series fed Hartley oscillator. Once a smooth, regulated dc source is obtained, most designers are reluctant to build circuits that introduce ripples back into the dc supply.

Otherwise, the operation of this circuit is exactly like that of the shunt fed Hartley oscillator. The feedback is developed between pins 1 and 2 of the transformer, and pin 1 is always 180 degrees out of phase with pin 3.

Push-pull Hartley Oscillator At times we encounter circuits that require more power than can be supplied by a single transistor oscillator. One way of solving this higher power problem is to modify the series fed Hartley oscillator into a push-pull oscillator using two transistors. Such an oscillator circuit is illustrated in Fig. 10-10.

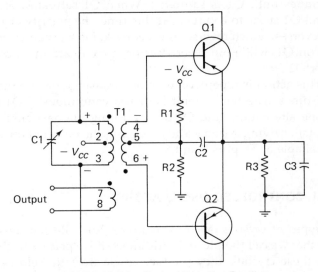

FIGURE 10-10. Push-pull Hartley Oscillator

Here we have two transistor amplifiers in common emitter configurations. The portion of the coil on T1 between pins 1 and 2 is the collector load impedance for Q1. The portion of the coil between pins 2 and 3 is the collector load impedance for Q2. Both transistors have common input bias developed across R1 and R2. Our transistors share a common circuit for their emitters, R3 and C3. Feedback for Q1 is developed between pins 4 and 5 of T1. Q2 gets its feedback from the voltage developed between pins 5 and 6 of T1. A third winding of the transformer, pins 7 and 8, is used to couple an output to another circuit.

Our two transistors should be matched as closely as possible, but a perfect match can never be achieved. When we apply power, both transistors start to conduct, but the slight mismatch causes one to conduct harder than the other. For the sake of explanation, let us suppose that Q1 is conducting more than Q2. In this case, Q1 collector current through the primary of T1 becomes dominant and energizes the coil with a positive at pin 1 and a negative at pin 3, as indicated in the illustration. This potential is magnetically coupled to the feedback coil, pins 4, 5, and 6. The initial feedback polarity is also indicated on the drawing.

The positive potential at T1 pin 6 opposes forward bias and stops the current in Q2. The negative potential at T1 pin 4 aids the forward bias on Q1 and aids the increasing current through this transistor. This condition, Q2 cut off and increasing current through Q1, continues until Q1 is saturated. When Q1 saturates, feedback ceases and Q2 starts to conduct. At this time, the polarity at T1 pins 1 and 3 reverses, which reverses the feedback. Pin 4 becomes positive and cuts off Q1, while pin 6 becomes negative to aid the increasing current of Q2.

This action is repetitive, with one transistor on and the other off at all times. The tank circuit is driven continuously. Q1 drives during one alternation, and Q2 drives during the next alternation. This constant driving action greatly increases the power of the output signal available at T1 pins 7 and 8.

CRYSTAL CONTROLLED OSCILLATORS

Certain types of crystals, such as quartz or Rochelle salt, possess a property that we call the piezoelectric effect. If we place a mechanical stress on these crystals, they produce an electrostatic field between the faces of the crystal. If we place an electrical potential between the two faces of the crystals, they produce a mechanical stress. Electrical energy produces a stress, the stress produces electrical energy, and the electrical energy produces a stress again. These actions are repetitive in alternate sequence.

The vibrations in the crystal closely resemble the oscillations in an LC tank circuit, and we frequently use these crystals instead of a tank circuit. The natural period of the mechanical vibrations is the frequency of the crystal, and each crystal is cut in a particular manner to oscillate at one specific frequency. When we use a crystal as the frequency determining device in an oscillator circuit, we may operate on the fundamental frequency of the crystal or at a frequency equal to any one of several harmonics of the fundamental frequency.

Circuit Characteristics A crystal holder is designed to protect the crystal and to bring an electrode into contact with each surface. Oscillations are very stable over a given temperature, and the holder may be designed to keep the crystal temperature within this range. Figure 10-11 illustrates the crystal in its holder and the equivalent circuit that this arrangement forms.

Figure 10-11(a) illustrates the crystal with its holder in symbolical form. Part (b) is the equivalent circuit of the crystal and its holder. C1 represents the parallel capacitance, and C2 represents the series capacitance. An inductor and a series resistance completes the equivalent components. C1 is the electrostatic capacitance between the crystal electrodes. C2 represents the series electrical equivalent of the mechanical vibrating characteristics of the crystal. The crystal has two separate frequencies of oscillation, as indicated in Fig. 10-12.

Frequency $f1$ represents the frequency where C2 and L form a series resonant circuit. Frequency $f2$ represents the frequency that

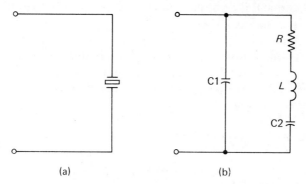

FIGURE 10-11. Crystal and Equivalent Circuit

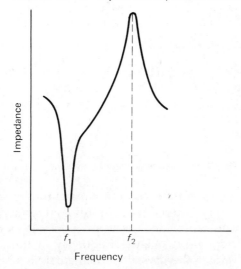

FIGURE 10-12. Impedance-frequency Curve

forms a parallel resonant circuit. At $f2$ the combined effects of L and C2 appear as a net inductor. C1 in parallel with this combination completes the parallel resonant tank circuit.

The graph in Fig. 10–12 *is not* a linear representation of the impedance-frequency relationship. It is intended only to show that the series resonant frequency point occurs at a point of relatively low impedance, whereas the impedance at parallel resonance is a great deal higher. Either series or parallel modes of oscillation may be used in an oscillator circuit. The mode of operation is determined primarily by the impedance of the circuit into which the crystal is connected.

The crystal acts as an extremely efficient tuned circuit and produces a remarkably steady frequency output. We can use crystals to control and stabilize any of the oscillators that we have discussed so far.

Crystal Controlled Armstrong Oscillator If we use a crystal as the control element (frequency control device) in our common emitter Armstrong oscillator, we shall have a circuit similar to that in Fig. 10–13.

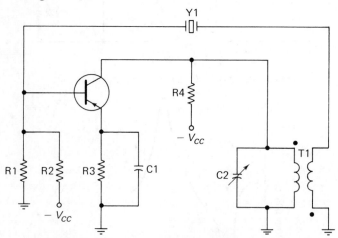

FIGURE 10–13. Crystal Controlled Armstrong Oscillator

Here we have a tuned collector Armstrong oscillator with a crystal serving as the control device. In this configuration, the crystal will operate in the series mode. The frequency is essentially fixed by the crystal frequency, but the frequency of the collector tank circuit may be either the crystal fundamental frequency or some convenient harmonic of this fundamental frequency. A variable frequency oscillator can be constructed by using interchangeable crystals. In this case, we would have a fixed crystal holder and several different crystals. C2 tunes the tank circuit and can vary the resonant frequency of the tank over a range of frequencies. The feedback is coupled across T1 to obtain the necessary phase shift. This feedback reinforces the crystal oscillations by replacing the energy lost in the circuit.

Crystal Controlled Colpitts Oscillator To provide crystal control for a Colpitts oscillator, we simply replace the transformer winding with a proper crystal. The identifying features of feedback are still present. Such a circuit is illustrated in Fig. 10–14.

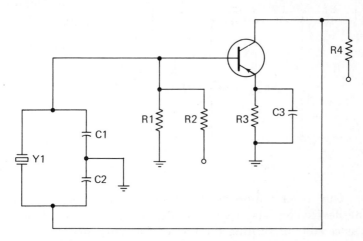

FIGURE 10-14. Crystal Controlled Colpitts Oscillator

Here our oscillating tank circuit consists of Y1, C1, and C2. C1 and C2 form an ac voltage divider for the feedback circuit. The voltage across C1 is the potential between base and emitter and is the feedback voltage. Except for the fact that the circuit is crystal controlled, it is obviously a Colpitts oscillator, and the function of all components is as previously described.

NONRESONANT FEEDBACK OSCILLATORS

So far we have discussed only oscillators with some type of resonant circuit. However, any amplifier will oscillate if it has a sufficient level of regenerative feedback, so it is possible to construct sine wave oscillator circuits without resonant networks. In this case, the feed-

254 Sinusoidal Oscillators

back circuit must supply the operating frequency as well as the proper phase shift.

Lag Line Oscillator The feedback circuit in our lag line oscillator is composed entirely of resistors and capacitors. This circuit is designed to provide the proper feedback and to determine the frequency of oscillation. Without the feedback network, this circuit is an ordinary amplifier. A circuit arrangement is illustrated in Fig. 10–15.

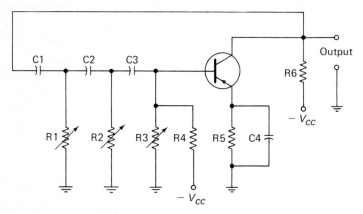

FIGURE 10–15. Lag Line Oscillator

Our combination feedback circuit and frequency determining device consists of C1R1, C2R2, and C3R3. A single RC network is capable of shifting the phase of a signal by a maximum of 90 degrees. We must have a phase shift of 180 degrees. This could possibly be accomplished by two capacitors and two resistors. However, we use three to provide room for adjustments. Each RC combination shifts the phase of the signal by about 60 degrees. Because of this stepping action in the phase shift, the lag line oscillator is also popularly known as a ladder oscillator and as a phase shift oscillator.

The oscillator is based on the principle that the RC network will provide exactly 180 degrees of phase shift for only one specific frequency. You know that the X_C of a capacitor is inversely proportional to the frequency. Now let us examine the combination of C1 and R1. When X_C and R are exactly equal, the phase shift is 45 degrees. If R is greater than X_C, the phase shift is less than 45 degrees. If X_C is greater than R, the phase shift is greater than 45 degrees. To accomplish the desired 60 degrees of shift, our X_C must be 1.5 times

R. For a given value of R, there is only one frequency that will produce that required value of X_C. The same reasoning applies to the other RC combinations.

Notice that our three resistors are variable. Changing the value of any given resistor requires a different frequency to provide the required 60 degrees of phase shift. Increasing resistance lowers the frequency. Decreasing the resistance raises the frequency.

Resistor R3 serves a dual function. It is one leg in the feedback circuit, and it teams up with R4 to form a voltage divider to control the input bias to the transistor. The output from this circuit is taken across R6.

The oscillations in this circuit are started by a surge of noise that is generated each time we apply power to the circuit. Noise frequencies range over the entire frequency spectrum, so one of these noise frequencies will be the same as the proper operating frequency of our circuit. This frequency will be shifted 180 degrees through the feedback circuit and become regenerative feedback. This frequency will be circulated and amplified over and over until it becomes the dominant frequency.

Energy losses in a lag line oscillator are relatively high. The high loss makes a high gain transistor essential to proper operation. The number of RC sections has a bearing on the circuit losses. If we use more sections, as a rule, losses decrease. So this circuit *is not* limited to only three RC sections.

Wien Bridge Oscillator The Wien bridge oscillator is another oscillator that uses RC networks to develop a sinusoidal output. It consists of two transistor amplifiers and an RC bridge. The RC bridge provides feedback and determines the frequency of oscillation. Such a circuit is illustrated in Fig. 10-16.

In this circuit, Q1 is our oscillator amplifier, and Q2 is an amplifier-inverter. The extra stage of amplification provides a higher output power while giving us a phase shift of 180 degrees for our feedback signal. Both of our transistors are in a common emitter configuration with unbypassed emitter resistors. The emitter resistor for Q1 is the thermistor; for Q2 it is R8. The degenerative feedback produced by these resistors provides our circuit with a high degree of frequency stability and reduces distortion of the signal. R3 in the bridge circuit also provides some degenerative feedback for the same purpose. We can afford to sacrifice some gain for increased stability, because we have two stages of amplification.

Our feedback is taken between the collector of Q2 and

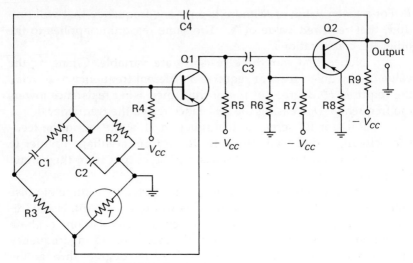

FIGURE 10-16. Wien Bridge Oscillator

ground. Trace this path through C4, and notice that this includes our entire bridge circuit. Positive feedback to the base of Q1 is developed across C1R1 and C2R2. When the feedback is at the operating frequency, there is no phase shift of the feedback signal. In this case, the feedback is in phase and aids the actions of Q1. At any other frequency, the feedback signal is shifted through the bridge and the regeneration is decreased.

A fixed portion of our feedback signal is developed across R3 and the thermistor. This is applied to the emitter of Q1 as degenerative feedback. Since this degenerative feedback is a fixed quantity and the regenerative feedback varies with our frequency, the degeneration is predominant for all off-frequency signals.

The thermistor between emitter and ground of Q1 provides us with a temperature control on the amplification of Q1. When our signal increases in amplitude, the current through our thermistor increases. The thermistor's resistance increases with current, because it has a positive temperature coefficient, and extra current raises its temperature. The increased resistance causes the negative feedback to increase and reduce the gain of Q1. The reduced gain brings our signal back to the proper amplitude. A decrease in signal amplitude has the opposite effect, and increases the gain of Q1.

Our frequency of operation is determined by the RC combinations of C1R1 and C2R2. We frequently find both C1 and C2 to be variable to provide a means of adjusting the frequency. In some cases, R1 and R2 are also variable for course frequency adjustments.

The operating frequency can be calculated from the values of RC by using this equation:

$$f = \frac{1}{2\pi \sqrt{R1 \times C1 \times R2 \times C2}}$$

When R1 is equal to R2 and C1 is equal to C2, we can simplify our equation, like this:

$$f = \frac{1}{2\pi\, R1C1}$$

REVIEW EXERCISES

1. Compare an electronic oscillator to a mechanical generator. Why do we need both?
2. Name the four essential sections of an oscillator.
3. How does the dc power supply for an oscillator differ from the supply for any other circuit?
4. List three characteristics of the common base configuration that are assets in an oscillator circuit.
5. List two characteristics of the common base configuration that are disadvantages in an oscillator circuit.
6. Why is the difference in input and output impedance a problem in an oscillator circuit?
7. List one disadvantage of using a common emitter configuration in an oscillator circuit.
8. State the impedance and gain characteristics for a common emitter configuration.
9. What are the primary disadvantages of using a common collector configuration as an oscillator circuit?
10. Define the flywheel effect.
11. Consider the drawing in Fig. 10–17. The capacitor is fully charged, and all current has ceased. Describe a complete cycle of operation when the switch is opened.

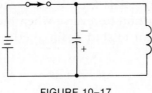

FIGURE 10–17.

258 Sinusoidal Oscillators

12. Suppose that we open the switch in the circuit of Fig. 10–17 and leave it open until all current ceases. Draw a waveshape to illustrate the complete chain of events in the tank circuit.
13. Indicate on the first cycle of item 12's waveshape where the current is being forced by the capacitor and where the inductor is the moving force.
14. What has to be done to a tank circuit to prevent the oscillations from dying out?
15. What determines the resonant frequency of an LC tank circuit?
16. What is the resonant frequency of the tank circuit in Fig. 10–18?

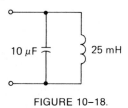

FIGURE 10–18.

17. How does the location of the operating point affect the frequency stability of an amplifier?
18. How does the temperature of the transistor affect the frequency stability?
19. At high frequencies, the collector to emitter capacitance becomes a problem. How can we minimize the effect of this capacitance?

Items 20 through 28 refer to the schematic in Fig. 10–19.

20. What function is performed by R1 and R2?
21. What does R3 do for the circuit?
22. What is the purpose of C2?
23. Describe the function of each of the three inductors.
24. The frequency of oscillation is determined by the values of what components?
25. How can we adjust the frequency of this oscillator?
26. What type of oscillator is this? What is the identifying feature?
27. What causes the action to reverse when the transistor saturates?
28. What causes current to start again after the transistor is driven to cutoff?

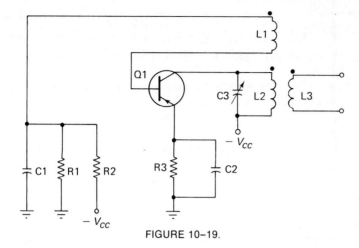

FIGURE 10-19.

29. What is the identifying feature of a Colpitts oscillator?
30. What is the principal difference between the Colpitts and Hartley oscillators? Explain.
31. What is the chief advantage of a push-pull oscillator over one that uses a single ended amplifier? Explain.
32. Define piezoelectric effect.
33. Why are crystals desirable for frequency control in oscillators?
34. What are the two modes of oscillation for a crystal?
35. What determines the mode of operation for a crystal?
36. Explain how a crystal of one frequency can control the frequency of a tank circuit that resonates at a higher frequency.
37. What provision is made in crystal controlled oscillators for varying the frequency?
38. Explain how an RC feedback network can determine the frequency of an oscillator.
39. What are two advantages of a Wien bridge oscillator?

Chapter Eleven

Nonsinusoidal Oscillators

In the previous chapter we discussed a variety of oscillators in the sinusoidal group; that is, oscillators that produce ac sine waves. We also have many applications of ac with waveshapes other than sine waves. There are a variety of oscillators that produce these nonsinusoidal ac waveshapes. We call this group of oscillators *nonsinusoidal* oscillators. These oscillators produce such signals as sawtooth sweep voltage, rectangular gating pulses, and sharp trigger pulses. We break down this group of nonsinusoidal oscillators into two main categories: *relaxation* oscillators and *synchronized* oscillators. The relaxation oscillators are free-running devices similar in operation to our sinusoidal oscillators. They require no input and supply a continuous string of output signals so long as we have power applied to them. The synchronized oscillators, on the other hand, produce an output only when they are triggered into action by an external signal. We shall discuss several examples of oscillators in each category.

RELAXATION OSCILLATORS

Most relaxation oscillators are based on the principle of transients in an RC circuit—the charge and discharge of a capacitor in series with a resistor. These RC circuits determine the frequency of oscillation. Let us briefly review this principle.

RC Time Constants When we connect a capacitor in series with a resistor and a dc source, at the first instant we have maximum current and maximum charging rate on the capacitor. As the capacitor charge increases, current decreases. When the capacitor charge equals the applied voltage, all current ceases. The time required to fully charge the capacitor depends upon the value of the resistor and the capacitor.

If we take a fully charged capacitor and connect it across a resistor, at the first instant we have maximum current and maximum discharging of the capacitor. As the capacitor charge decreases, current decreases. When the capacitor is fully discharged, all current ceases. The time required to fully discharge the capacitor is also dependent upon the value of the resistor and the capacitor. If we have the same capacitor, the same resistor, and the same voltage, the charge time is exactly the same as the discharge time. Consider the illustration in Fig. 11-1.

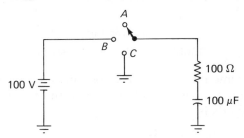

FIGURE 11-1. Charge and Discharge of a Capacitor

The instant that we move the switch from A to B, we complete the circuit and have maximum current from ground to the bottom plate of the capacitor and from the top plate through the resistor. The current gradually decreases as the capacitor charges with positive at the top plate and negative on the grounded plate. When the capacitor has charged to 100 V, its charge is equal and opposite to the applied voltage, and all current ceases.

Now if we move the switch from B to C, we remove the battery from the circuit and complete a path for the capacitor to discharge through the resistor. At the first instant, we have the full 100-V charge driving the current, so we have maximum current. As the charge gradually decreases, the current grows smaller. The current ceases when the capacitor is fully discharged. The time for the discharge is exactly the same as the time for the charge. For an exact picture of the time for charge and discharge, examine the graph in Fig. 11-2.

The quantity of voltage applied has no bearing on the time; time is solely a function of the value of resistance and capacitance. We define a time constant as the time required for a capacitor to charge to 63 percent of the value of the applied voltage. Since it dis-

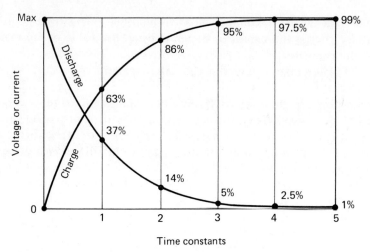

FIGURE 11-2. RC Time Constants

charges at the same rate, we can also say that a time constant is the time required for a capacitor to lose 63 percent of its charge. A time constant in seconds is

$$t = RC$$

For our circuit in Fig. 11-1, one time constant is

$$t = RC$$
$$= 100 \text{ } \Omega \times 100 \text{ } \mu F$$
$$= 10{,}000 \times 10^{-6}$$
$$= 0.01 \text{ s}$$

Notice on the graph in Fig. 11-2 that five time constants are required for either charge or discharge. The percentage figures indicate the quantity of charge on the capacitor at the end of each time constant. About 99 percent of the applied voltage is on the capacitor after five time constants; we consider this to be fully charged. About 1 percent of the total charge is still on the capacitor after it has been discharging for five time constants; we consider that this is fully discharged. For the circuit in Fig. 11-1, the capacitor will charge in 0.05 s and discharge in 0.05 s:

$$t = 5RC$$
$$= 5 \times 100 \text{ } \Omega \times 100 \times 10^{-6}$$
$$= 50{,}000 \times 10^{-6}$$
$$= 0.05 \text{ s}$$

Most solid state relaxation oscillators are composed of RC circuits that use some type of solid state device as an electronic switch. The opening and closing of the switch determines when the capacitor will charge and when it will discharge.

Simple Relaxation Oscillators Building on our circuit of Fig. 11-1, if we replace the mechanical switch with an electronic switching device, we will have the simplest form of relaxation oscillator. Using a unijunction transistor as a switching device produces the circuit in Fig. 11-3.

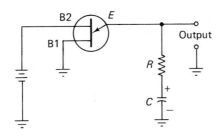

FIGURE 11-3. Unijunction Relaxation Oscillator

In this arrangement, we have the full battery voltage between the two bases with negative at B1 and positive at B2. The silicon bar between the two bases acts as a voltage divider. There is a voltage gradient in the bar causing the emitter potential to fall some place between these two extremes.

When power is applied, both junctions are reverse-biased. A small reverse current from emitter to B2 begins to charge the capacitor with the indicated polarity. As the capacitor's charge increases, the emitter becomes more positive. When the emitter potential reaches a certain level, more positive than B1, that junction becomes forward-biased. In effect this closes the switch from B1 to the emitter and provides a discharge path for the capacitor. Electron current from the negative plate of the capacitor passes from ground to B1, to the emitter, through the resistor, and back to the top plate of the capacitor.

As the capacitor discharges, the potential on the emitter grows smaller. But the resistance between B1 and the emitter has become very nearly a short, and the emitter potential drops very low before the cutoff point is reached. At cutoff, the switch between B1 and the emitter is open, and current ceases. Reverse current from emitter to B2 again starts to charge the capacitor in the original direction. Figure 11-4 illustrates a waveshape from this oscillator.

Notice that the charge and discharge curves closely resemble

264 Nonsinusoidal Oscillators

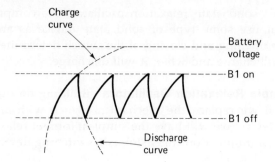

FIGURE 11-4. Output from Unijunction Oscillator

our curves in Fig. 11-2. Charge time and discharge time both depend upon the value of R and C, so the values of R and C determine the frequency and shape of the sawtooth output. Such waveshapes can be used for a variety of timing applications, and the time for each cycle can be made adjustable by using either a variable resistor or a variable capacitor.

If we rearrange our circuit so that the charge path is through the resistor and our capacitor is shorted for quick discharge, we have all the principles of a free-running sweep generator. We find a variety of solid state devices used for the automatic switching action. Such an oscillator is illustrated in Fig. 11-5.

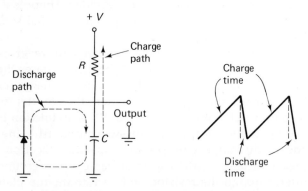

FIGURE 11-5. Zener Diode Relaxation Oscillator

This is one of our simplest relaxation oscillators, and at the same time, our simplest form of a sweep generator. When power is applied, electrons flow from ground to the bottom plate of the capacitor, and from the top plate of the capacitor to the positive voltage supply. The potential at the junction of R and C grows more positive at an exponential rate. When this potential reaches the breakdown potential of the Zener diode, the avalanche current through the Zener shorts this point to ground. The capacitor then rapidly discharges through the Zener diode. Once the capacitor has discharged, the potential at the junction is effectively zero and causes avalanche current to cease. The Zener is now an open switch again, and the capacitor starts to charge again. This action continues so long as we have power to the circuit.

Notice on our sawtooth output that we have a slow charge time and a very rapid discharge time. The time difference is caused from the different resistance values in the charge and discharge paths. A waveshape similar to this is used to move the electron stream across the face of an oscilloscope to draw the time base line. During charge time, the sweep moves from left to right across the screen. During discharge time, the sweep moves rapidly back to the left of the screen in preparation for the start of another sweep. In this application, we call the discharge time *flyback* time. The trace is usually shut off during flyback time to produce a cleaner picture.

Notice also that the charge time is very nearly linear. How did we get from an exponential rise to a linear rise? Go back to the time constant graph. Our capacitor charges to 63 percent of the applied voltage during the first time constant. Now suppose that we allow the capacitor to charge only about 10 percent before we close the switch and cause it to discharge. The first 10 percent of a capacitor's charge along this curve is very nearly a linear rise. That 10 percent is what we use for a sweep voltage.

A capacitor will charge to 10 percent of the applied voltage in about 0.08 TC. Suppose that our circuit in Fig. 11–5 has a resistance of 200 kΩ and a capacitance of 30 μF. What is the time for each linear sweep?

$$1\text{TC} = RC$$
$$= 200 \times 10^3 \times 30 \times 10^{-6}$$
$$= 6 \text{ s or } 6000 \text{ ms}$$
$$\text{Sweep time} = 0.08 \text{ TC}$$
$$= 0.08 \times 6000$$
$$= 480 \text{ ms}$$

Astable Multivibrator The astable multivibrator is another type of free-running (relaxation) oscillator. We find it widely used to generate rectangular or square waves for gating and timing operations. The astable multivibrator consists of two amplifiers with the output of each RC coupled to the input of the other. One circuit arrangement for this oscillator is illustrated in Fig. 11-6.

Notice that we have two NPN transistors, each in a common emitter configuration. The two amplifiers and the circuit components are matched as nearly as possible. The potential at point A causes C1 to charge and discharge to control the bias on Q2. The potential at point C causes C2 to charge and discharge to control the conduction of Q1.

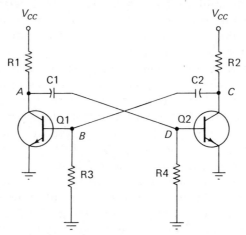

FIGURE 11-6. Astable Multivibrator

When we apply power, both transistors start to conduct, but because of a slight mismatch, one transistor will conduct more than the other. The voltage at the collector of the dominant transistor drops. This drop in potential causes the coupling capacitor to start discharging through the base resistor of the other transistor. The resulting negative potential across the base resistor cuts off the slow transistor and holds it cut off until the capacitor charge is nearly dissipated. Now the cutoff transistor starts to conduct, and the feedback cuts off the first transistor. This action is repetitive, with each transistor being on half the time and off half the time. The drawing in Fig. 11-7 shows the waveshapes at points A, B, C, and D of the schematic.

Let us pick up the action with the circuit already in operation. At $t1$, transistor Q1 has just started to conduct. This is evident from the fact that the waveshape on its base, point B, is at a level above cutoff. As Q1 starts to conduct, the collector voltage, point A, drops. This drop in potential forces C1 to start discharging through R4 to ground. This discharge current places a strong negative potential at point D, which cuts off Q2. The potential at the collector of Q2

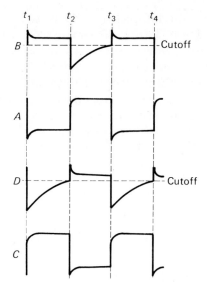

FIGURE 11-7. Astable Waveshapes

goes to the level of V_{CC} when current through R2 ceases. C2 starts to charge toward V_{CC} by drawing current from ground through R3. This charging current holds a positive potential at point B and keeps Q1 conducting at saturation.

The described conditions hold until $t2$. During the time from $t1$ to $t2$, C1 was gradually losing its charge and the potential at point D was gradually rising. At $t2$, C1 has discharged to a level that allows Q2 to conduct. The collector of Q2 drops from V_{CC} to nearly zero as Q2 goes to saturation. This low potential at point C forces C2 to start discharging through R3 to ground. This discharge current through R3 holds a negative potential on the base (point B) of Q1 and cuts it off.

Transistor Q1 will stay cut off and Q2 will remain saturated until $t3$. At $t3$, C2 has lost nearly all its charge and the potential at point B allows Q1 to start conducting again. This is a repeat of the conditions at $t1$. Q1 goes to saturation and Q2 goes to cutoff. The time from $t1$ to $t2$ is determined by the RC time constant of C1R4. The time for the other half-cycle, $t2$ to $t3$, is determined by C2R3. The outputs are normally taken from the two collectors. Waveshape A is available at the collector of Q1. Waveshape C is available at the collector of Q2. The two outputs are identical, except that they are 180 degrees out of phase.

Blocking Oscillator The free-run blocking oscillator is another type of self-starting (relaxation) oscillator. This circuit uses a class C amplifier and produces trigger pulses of short duration. The relatively simple circuit of this oscillator is illustrated in Fig. 11–8.

When we apply power, a low current starts through the transistor. The collector current causes a decrease in potential at point A. This decrease is transformer-coupled to point B as a positive potential. The positive at point B causes the capacitor to start charging by drawing current from ground through R1. The current through R1 places an increasing positive potential at point C. The action is accumulative and rapidly drives our transistor to saturation. When saturation is reached, feedback ceases, and C1 starts to discharge through R1 to ground. The current through R1 places a strong negative potential at point C, which cuts off the current through the transistor. Our transistor will be held cut off until the capacitor is fully discharged. The cutoff time will be five time constants of R1C1. At the end of this time, the transistor starts conducting again and we start another cycle.

Waveshape D is the output signal at point D. Taking it across this transformer, we have a 180-degree phase shift from point A, and this gives us a positive trigger pulse for our output. These pulses will be very short in duration and will be separated by a time equal to 5RC.

Waveshape C appears at point C: the base of our transistor. It shows the charge and discharge of capacitor C1.

Since frequency is equal to the reciprocal of time, let us assume values and calculate time and frequency. Suppose that R1 = 50 Ω and C1 = 10 μF. We can calculate the time between pulses as follows:

$$t = 5RC$$
$$= 5 \times 50 \times 10 \times 10^{-6}$$
$$= 2.5 \times 10^{-3}$$
$$= 2.5 \text{ ms}$$

Frequency then is

$$f = \frac{1}{t}$$
$$= \frac{1}{2.5 \times 10^{-3}}$$
$$= 0.4 \times 10^3$$
$$= 400 \text{ Hz}$$

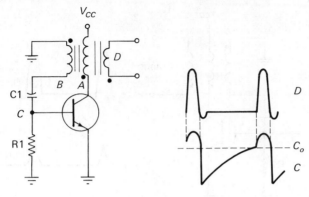

FIGURE 11-8. Blocking Oscillator

With pulse waveshapes, we often refer to the frequency as pulses per second or pulse recurrence frequency (PRF) instead of hertz. We can alter the time, and so change the PRF, by changing the value of either R1 or C1.

SYNCHRONIZED OSCILLATORS

All of the relaxation oscillators that we previously discussed can be modified into synchronized oscillators. A synchronized oscillator produces an output each time it is triggered by a pulse or a set of pulses from an external source. We have a great many oscillators that fit into this category. The monostable multivibrator, the bistable multivibrator, the Schmitt trigger, the synchronized sweep generator, and the synchronized blocking oscillator are all examples of synchronized oscillators.

Synchronized Blocking Oscillator Only a slight circuit change is required to convert our free-run blocking oscillator into a synchronized oscillator. We need two changes: cutoff bias for the amplifier and provision for an input trigger pulse. Such a circuit is illustrated in Fig. 11-9.

With cutoff bias as illustrated, our transistor cannot conduct until a positive input trigger overcomes the bias. Each positive input triggers the transistor into conduction. It then goes through one cycle; it goes to saturation, then back to cutoff, and produces one output pulse. The $-V_{BB}$ on the base holds it cut off until another positive pulse is applied. The PRF of the output is the same as that of the input.

270 Nonsinusoidal Oscillators

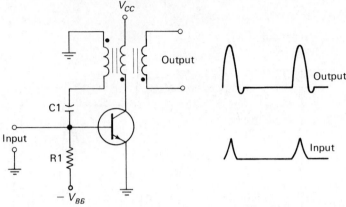

FIGURE 11-9. Synchronized Blocking Oscillator

The natural rest period for this oscillator is 5R1C1, and this determines the normal free-run frequency. We may remove the negative voltage from the base, replace it with ground, and still have a synchronized blocking oscillator under one condition. The frequency (PRF) of the input signal would have to be slightly higher than the natural free-run frequency of the oscillator. Under these circumstances, the positive trigger for each cycle would arrive just before C1 was completely discharged. This would trigger the action just before the transistor started to conduct of its own volition.

Monostable Multivibrator Another of our widely used synchronized oscillators is the monostable multivibrator. It produces a single rectangular output with each input trigger pulse. The duration, amplitude, and shape of the output depend upon the value of the circuit components, whereas the frequency is determined by the frequency of the input. An example of this circuit is shown in Fig. 11-10.

In the quiescent condition, transistor Q1 is biased to cutoff by $-V_{BB}$ through R5 to its base. During this time, the collector voltage of Q1 is the same as V_{CC}, because we have zero current through R1. The positive potential on this collector charges C1 with the polarity indicated. V_{CC} through R2 holds a positive potential on the base of Q2 and keeps it conducting at saturation. The high current through R3 keeps the collector voltage of Q2 at a low positive potential: effectively zero voltage. This is a stable state, and it will continue until an external positive pulse through C1 overcomes the bias of Q1 and triggers it into conduction.

As we apply an external positive pulse through C2, Q1 is

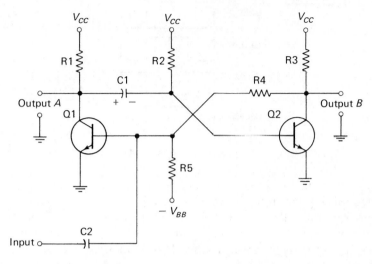

FIGURE 11-10. Monostable Multivibrator

triggered into conduction. Current through R1 lowers the positive potential at the collector of Q1 and causes C1 to start discharging through R2. Current through R2 places a strong negative potential on the base of Q2 and drives it from saturation to cutoff. The absence of current through R3 allows the collector of Q2 to rise to the level of V_{CC}. This positive potential is felt across R4 to the base of Q1, and it drives Q1 into saturation. This is the beginning of our unstable state.

The unstable state continues as C1 discharges through R2. As C1 discharges, the potential on the base of Q2 rises. After a time, this potential reaches a level that permits Q2 to conduct again. When Q2 starts to conduct, the current through R3 causes Q2's collector voltage to drop. This drop in potential is felt through R4 on the base of Q1. Q1 conducts less, and the collector voltage rises. This rise in voltage stops the discharge of C1 and removes the negative from the base of Q2. V_{CC} through R2 to the base of Q2 now drives Q2 into saturation. The large current through R3 drops Q2's collector voltage to zero. This zero potential felt across R4 allows $-V_{BB}$ to take control of Q1 and return it to cutoff status. This ends the unstable state and places the circuit back into a stable state. The stable state will continue until we apply another positive pulse through C2 to trigger another cycle of events. Figure 11-11 illustrates the waveshapes for this circuit.

The output may be taken from the collector of either (or both) transistor. These two waveshapes are always 180 degrees out of

272 Nonsinusoidal Oscillators

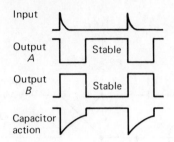

FIGURE 11–11. Waveshapes for Monostable Multivibrator

phase. During the unstable state, we have a choice of a negative rectangular wave from the collector of Q1 or a positive rectangular wave from the collector of Q2. Both output A and output B start with the input signal and terminate when C1 discharges. The frequency of both outputs is the same as that of the input trigger. Since we have one output each time the oscillator is triggered, we sometimes call this circuit a *one-shot* multivibrator.

One application of the monostable multivibrator is to provide a delayed trigger for some other circuit. In this case, the output is differentiated, and we use the trigger that is created by the trailing edge of the output rectangular wave. In effect, this delays the input trigger for a period of time equal to the unstable state (output duration) of our oscillator. The duration of the output can be changed by using a different value for either C1 or R2. The output duration can be made variable by using a variable component for either C1 or R2.

Bistable Multivibrator The bistable multivibrator is a very popular and useful synchronized oscillator. This circuit uses two transistors in a balanced arrangement that provides two stable states. It requires a trigger pulse from an external source in order to switch from either of the stable states to the other stable state. A circuit and waveshapes for this oscillator are illustrated in Fig. 11–12.

When we apply power, both transistors start to conduct. Since a perfect match is impossible to achieve, one transistor conducts more than the other and becomes dominant. The dominant transistor goes to saturation and the other goes to cutoff. This is a stable state, and it will hold until a positive trigger is applied to the base of the cutoff transistor. Each positive pulse triggers a cycle of events that changes the saturated transistor to cutoff and the cutoff transistor to saturation.

The waveshapes in Fig. 11–12 show a condition where Q1 was saturated and Q2 was cut off prior to the arrival of the input

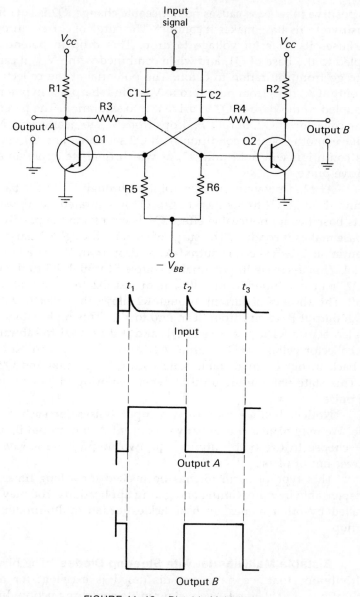

FIGURE 11-12. Bistable Multivibrator

trigger at $t1$. Output A, the collector of Q1, was at a very low potential. Output B, the collector of Q2, was at a high potential.

At $t1$, a positive trigger pulse is applied through C1 to the base of Q2, and through C2 to the base of Q1. Q1 is already saturated, so a positive to its base causes no noticeable change. Q2 is cut off, so a positive to its base makes it conduct. The surge of current through Q2 causes its collector voltage to drop. This drop in potential is coupled to the base of Q1, and when combined with $-V_{BB}$, it causes Q1 to go from saturation to cutoff. The potential at the collector of Q1, output A, rises from near zero to V_{CC}. This sharp rise in potential is coupled to the base of Q2 to drive it into saturation. The potential at the collector of Q2, output B, drops from V_{CC} to nearly zero. Now we have another stable condition with Q2 saturated and Q1 cut off. This condition will hold until $t2$, at which time we apply another positive pulse.

At $t2$, a positive pulse is applied through C1 to the base of Q2 and through C2 to the base of Q1. Q2 is saturated, so a positive on its base has no noticeable effect. Q1 is cut off, so the positive on its base makes it conduct. The surge of current through R1 causes the potential at Q1's collector, output A, to drop from V_{CC} to a low potential. This decrease in potential is coupled through R3 to the base of Q2, and combined with $-V_{BB}$, it drives Q2 from saturation to cutoff. The absence of current through R2 allows the collector voltage of Q2, output B, to rise from near zero to V_{CC}. This high potential is coupled across R4 to the base of Q1, and it drives Q1 to saturation. The collector voltage of Q1, output A, drops to nearly zero. Now we are back in our original stable state with Q1 saturated and Q2 cut off. This state will endure until $t3$, when we bring in another positive pulse.

Notice that we require two input pulses for each output cycle. We may have a choice between output A and output B, or we may choose to use both outputs. The two outputs are always 180 degrees out of phase.

This type of oscillator has been used for a long time. It is indispensable in many timing and gating applications. You may find it called by other names, such as Eckles-Jordan multivibrator and flip flop.

Bistable Multivibrator with Steering Diodes The bistable multivibrator that we have just discussed is excellent for many applications, but it is relatively slow in its switching action. In circuits that require a high degree of timing precision, we might require a faster switching oscillator. There are two modifications that we can make on the bistable multivibrator to increase its switching speed: We can add bypass capacitors across the coupling resistors, and use diodes to steer the input trigger pulses. Figure 11–13 shows a bistable multivibrator that has these two modifications.

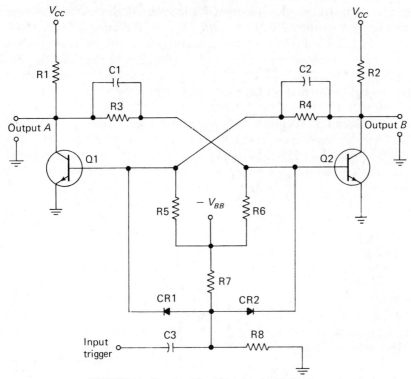

FIGURE 11-13. Modified Bistable Multivibrator

Capacitors C1 and C2 are coupling capacitors that provide a very fast reaction in the feedback circuits. A change at one of the collectors is instantly delivered to the base of the other transistor.

The steering circuit consists of CR1, CR2, R7, and R8. This steering circuit guides the positive input trigger pulses so that they are always applied only to the cutoff transistor. Let us assume a condition when Q1 is saturated and Q2 is cut off.

We need to apply the next positive pulse to the base of Q2 only. The bias on CR1 is a very high reverse bias, consisting of V_{CC} on the cathode and $-V_{BB}$ on the anode. This diode cannot possibly conduct under these bias conditions. The bias on CR2 is very near zero at this time; it has a small negative potential on the anode and nearly zero potential on the cathode. The next positive trigger pulse that couples through C3 will cause CR2 to conduct, and this couples the pulse to the base of Q2, the cutoff transistor. This pulse causes the multivibrator to change states.

Now we have the reversed condition. Q1 is cut off and Q2 is saturated. We need to apply the next positive trigger to the base of Q1. At this time, the bias on CR2 is a very high reverse bias consisting of V_{CC} on the cathode and $-V_{BB}$ on the anode. CR2 cannot conduct. But the bias on CR1 has decreased to very near zero; it has a small negative on the anode and nearly zero on the cathode. The next positive trigger through C3 causes CR1 to conduct and couple the signal to the base of Q1, our cutoff transistor.

Schmitt Trigger The Schmitt trigger is a synchronized oscillator that will convert almost any type of signal into a rectangular waveshape. The only requirements for the input signal are proper polarity and amplitude to activate the circuit. A schematic of the Schmitt trigger along with waveshapes is illustrated in Fig. 11-14.

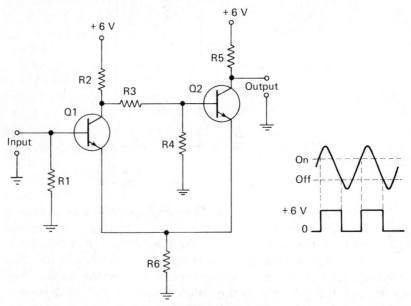

FIGURE 11-14. Schmitt Trigger

When we apply power, both transistors start to conduct, but Q2 quickly goes to saturation while Q1 cuts off. This initial action is caused from the circuit arrangement. R2, R3, and R4 form a voltage divider that reduces the collector voltage on Q1 while placing a small positive (forward bias) on the base of Q2. Current for both transistors comes from ground through R6 and develops a positive potential on both emitters. Q2 is conducting too hard to be affected, but this positive potential places reverse bias on Q1 and cuts it off.

The output is taken from the collector of Q2. With Q2 saturated, the output is zero; with Q2 cut off, the output is +6 V. Any input signal that will turn Q1 on and off will produce a square or rectangular pulse from Q2. The +6 V used for V_{CC} will produce output amplitudes from 0 to +6 V; a larger voltage will produce an output of greater amplitude.

In the illustration, we have assumed a sine wave input. This sine wave is applied to the base of Q1. Notice on the waveshape that different levels are designated as on and off. As the sine wave rises in a positive direction, it reaches a point where Q1 is turned on. When Q1 conducts, the collector voltage drops to a very low value and greatly reduces the positive potential on the base of Q2. Reduced voltage at the base, combined with the positive on the emitter, causes Q2 to cut off. When Q2 goes from saturation to cutoff, the collector potential, our output, rises from zero to +6 V and holds this level as long as Q2 is cut off. The input rises to peak value and starts to decrease, but it passes through zero before it causes Q1 to cut off. When the sine wave is negative enough to cut off Q1, the collector voltage rises to almost +6 V and causes Q2 to conduct again. Q2 goes from cutoff to saturation in a very short time. The collector voltage of Q2, our output, drops from +6 V to zero.

The Schmitt trigger is primarily employed in wave shaping operations. With a sine wave input as we have here, it will produce a string of uniform rectangular waves. But as we mentioned before, we are not restricted to any particular type of input. Any waveshape that will cause Q1 to conduct will produce a square or rectangular wave output.

Synchronized Sweep Generator The synchronized sweep generator is a synchronized sawtooth oscillator that is very valuable in many types of timing circuits. It is particularly useful in producing the time base for oscilloscopes. This circuit differs only slightly from our free-run sweep generator, as you can see by the schematic in Fig. 11-15.

This circuit arrangement is actually a free-run sweep generator, and it will free-run in the absence of an input trigger. This means that the frequency (PRF) of the trigger pulses must be just a little higher than the natural free-run frequency of the oscillator.

When we apply power, the capacitor starts to charge by current along the indicated path. As it charges, the collector voltage of the transistor rises more and more positive. Without an input signal, this rising potential will reach a level that will cause the transistor to conduct. The conducting transistor effectively shorts out the capacitor and provides a quick discharge path. This would

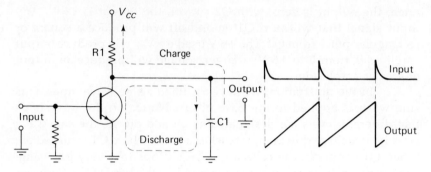

FIGURE 11-15. Synchronized Sweep Generator

be a free-run sweep generator. We do not want it to free-run, so we send in a positive trigger pulse to turn on the transistor just a little early. The trigger pulse causes Q1 to conduct and terminates the charge on C1. Since another sweep starts immediately after each sweep terminates, our trigger pulse actually does two jobs: It terminates the sweep that is in progress and starts the next sweep. The sweep time is the same as the time between trigger pulses, and this makes the input frequency (PRF) equal to the output frequency.

REVIEW EXERCISES

1. What is the primary difference between a sinusoidal oscillator and a nonsinusoidal oscillator?
2. Name the two main categories of nonsinusoidal oscillators.
3. Describe the identifying features of oscillators in each category of item 2.
4. What determines the frequency of oscillation for a relaxation oscillator?

Items 5 through 10 refer to the illustration in Fig. 11-16.

5. In terms of RC time constants, how long will it take to charge the capacitor to 10 V after the switch is moved to position 2?
6. Convert your answer to item 5 to seconds.
7. Assume that the capacitor is fully charged and the switch is moved to position 3. In terms of time constants, how long will it take the capacitor to discharge to

   a. 3.7 V?
   b. 0.5 V?
   c. 0 V?

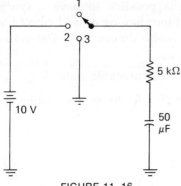

FIGURE 11–16.

8. Convert your answers to item 7 into seconds.
9. Assume that the switch is moved to position 2 for one time constant and then moved to position 3.

   a. What was the maximum charge acquired by the capacitor?
   b. How many time constants are required to reduce this charge to zero?

10. Compare the action of this circuit and the action in a simple relaxation oscillator.
11. Draw a schematic of a simple Zener relaxation oscillator.
12. What type of waveshape is produced by the oscillator of item 11?
13. With regard to the oscillator of item 11, what determines when the capacitor will charge and discharge?
14. When one is using a capacitor charge as a sweep voltage, what percentage of the charge will produce a linear sweep?
15. What causes an astable multivibrator to start oscillating?
16. What determines the frequency of the astable multivibrator?
17. What type of signals are produced by the astable multivibrator, and where is the output taken?
18. What kind of signals are produced by a free-run blocking oscillator?
19. What causes the transistor of a free-run blocking oscillator to

    a. Start conducting?
    b. Go to saturation?
    c. Cut off?

20. Normally, what two changes would we make to convert a free-run blocking oscillator to a synchronized blocking oscillator?
21. What is the frequency relationship of the input to output of a synchronized blocking oscillator?

280   Nonsinusoidal Oscillators

22. Explain how it is possible to have a synchronized blocking oscillator without having a fixed cutoff bias.
23. In a monostable multivibrator, what determines the
    a. Frequency?
    b. Duration of the unstable state?

Items 24 through 29 refer to the illustration in Fig. 11–17.

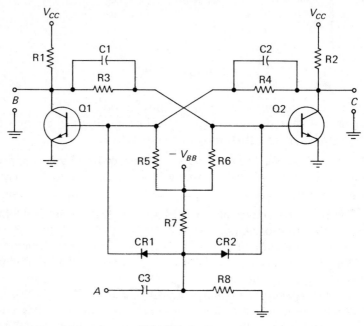

FIGURE 11–17.

24. Identify the circuit.
25. What is the advantage of this circuit over the conventional bistable multivibrator?
26. What function is served by the diodes and their associated circuits?
27. What function is served by C1 and C2?
28. Draw the waveshapes for one output cycle at points A, B, and C, showing the time relationship. Start with Q1 saturated.
29. What is the frequency relationship between the input and output?
30. Describe the input and output of a Schmitt trigger.

Chapter Twelve

Modulators

Electromagnetic waves are used to establish communication links throughout the world. With the use of space probes, we have now extended these links to include most of our solar system. We sometimes refer to these electromagnetic waves as radio waves or carrier waves. These waves carry many kinds of intelligence, such as voice, music, teletype, video, facsimile, and telemetry, from the transmitter to the receiver. The process of impressing intelligence upon a carrier (electromagnetic) wave is called modulation. The device that we use for making this impression is called a modulator. We shall discuss the principles of modulation as well as several types of modulators.

There are two principal types of modulation in use today. They are amplitude modulation and frequency modulation. The principles of these two types are different and the modulators are different. We shall discuss both amplitude and frequency modulation, along with the modulators used to accomplish each type of modulation.

MICROPHONES

You are right. A microphone *is not* a modulator. But microphones, transducers, and other related devices are used to convert intelligence into electrical energy. Since these devices are essential to the transmission of intelligence, we shall discuss a few types of microphones.

A microphone is a device that converts sound energy to electrical energy. The sound waves cause compression and rarefaction of the atmosphere, which will cause the diaphragm of a microphone to move in and out. The diaphragm is connected to some device that causes current flow in proportion to the instantaneous pressure delivered to it. We have many types of microphones. The one that we select to use in any particular situation depends upon what the application demands, such as sensitivity, frequency response, impedance, power requirements, and ruggedness.

Carbon Microphone The carbon microphone is based on the principle that the resistance of a carbon pile will vary with the pressure applied to the carbon granules. The diaphragm is arranged to vary the pressure as the diaphragm vibrates. Since the vibrations are caused by sound waves, the sound waves indirectly control the resistance of the carbon pile. A microphone circuit is illustrated in Fig. 12-1.

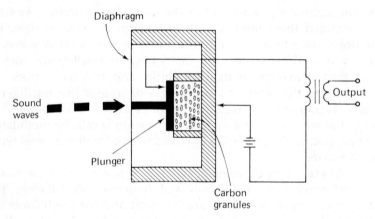

FIGURE 12-1. Carbon Microphone

The battery sends a direct current through the carbon pile and through the transformer. The level of current depends upon the resistance of the carbon pile. This resistance, in turn, depends upon the pressure applied to the carbon granules by the plunger. The sound waves striking the diaphragm cause it to vibrate and vary the plunger with the vibrations. These vibrations vary the resistance and consequently vary the current in the circuit. The varying level of direct current is coupled across the transformer as alternating current. The output is generally coupled to an audio amplifier. The transformer matches the low impedance of the microphone output to the high impedance of the amplifier input and steps up the voltage in the process. After proper amplification, this output will be used as a modulating wave for a carrier wave. Such a carbon microphone is lightweight, rugged, and has high output. However, its fidelity is poor. Random changes in the resistance of the carbon granules tend to produce an undesirable background hiss. Also, the granules may pack together and cause reduced sensitivity and distortion. It has a limited frequency response, which makes it acceptable for voice communications but undesirable for music.

Crystal Microphone Our crystal microphone is delicate and sensitive, but it has few of the disadvantages of the carbon microphone. It utilizes the piezoelectric qualities of Rochelle salt or quartz crystals and generates its own voltage. It has a high impedance output and can be connected directly into an amplifier stage. The output is very low and may require several stages of amplification before the signal is used to modulate a carrier wave. One form of crystal microphone is illustrated in Fig. 12-2.

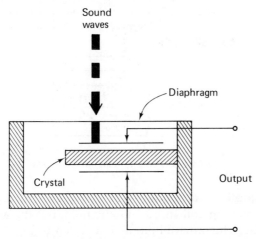

FIGURE 12-2. Crystal Microphone

As the sound waves strike the diaphragm, a vibration is set up. This vibration is transferred to the crystal. The resulting mechanical stress causes the crystal to produce a potential between its two surfaces. The potential varies with the sound waves. This varying potential output is coupled to an amplifier and handled as any weak ac signal.

Dynamic Microphone Our dynamic microphone is based on the principle of a moving coil in a permanent magnetic field. A coil of fine wire is mounted on the back of the diaphragm. This coil is in a magnetic field and moves when the diaphragm vibrates. We have an alternating current induced into the moving coil, which is the electrical equivalent of the sound waves striking the diaphragm. This microphone is illustrated in Fig. 12-3.

The dynamic microphone has a reasonable degree of sensitivity, is lightweight, and is rugged. It can withstand vibration, temperature variations, and moisture better than most other microphones. It has a uniform frequency response at frequencies from 40 to 15,000 Hz. The output impedance of the dynamic microphone is very low, and we nearly always couple the output through a transformer to match the impedance to the amplifier input.

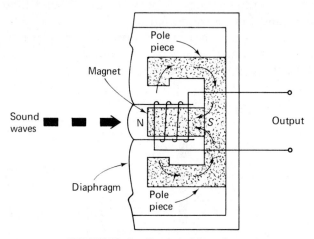

FIGURE 12-3. Dynamic Microphone

Magnetic Microphone Our magnetic microphone is also based on a moving coil in a magnetic field, but the arrangement is rather unique. This microphone is illustrated in Fig. 12-4.

The arrows in this illustration show the direction of our magnetic flux through the pole pieces and through the coil. Notice that the flux through the coil can go either way. As sound waves strike the diaphragm, the diaphragm vibrates and moves the plunger in and out. The plunger is mechanically coupled to a flexible magnetic rod, and it moves this rod in the air gap between the pole pieces. In a static condition, there is no flux through the magnetic rod or through the coil. A compression wave pushes against the diaphragm and moves the magnetic rod to the right near pole piece A. Some of the flux now detours down through the iron rod and generates a current in the coil. A rarefaction wave causes the diaphragm to bow outward and pull the magnetic rod to the left near pole piece B. Some of the magnetic flux now moves upward through the rod and generates a

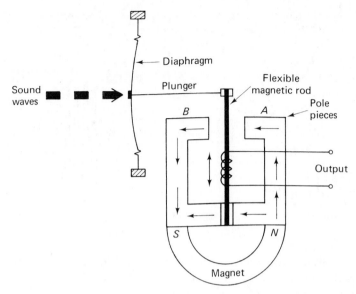

FIGURE 12-4. Magnetic Microphone

current in the coil. Each time the flux reverses, the polarity of the generated current reverses. The quantity of flux through the rod and, therefore, the amplitude of the generated current are determined by the volume of the sound. The vibrations of the diaphragm cause an alternating flux, which induces an alternating output current.

The magnetic microphone is more resistant to rough handling than most other microphones. It can withstand considerable vibration and shock. It has a relatively high sensitivity and a frequency response from 40 to about 15,000 Hz. Since our output is from a coil, the impedance is very low. We nearly always couple the output through a transformer in order to match the impedance to the amplifier stage.

AMPLITUDE MODULATION PRINCIPLES

Amplitude modulation is a process of combining an intelligence signal, such as the output of a microphone, with a carrier wave. The carrier wave retains its frequency, but its amplitude is caused to vary at the frequency of the intelligence signal. We refer to this intelligence signal as the modulating wave. Our intelligence is impressed into the amplitude of the carrier wave. This principle is illustrated in Fig. 12-5.

286 Modulators

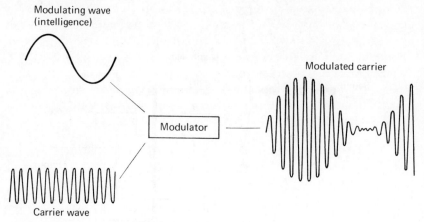

FIGURE 12-5. Amplitude Modulation

Sidebands When we mix two signals as described, we generate two new frequencies: the *sum* of the two frequencies and the *difference* between the two frequencies. These two new frequencies both carry the complete intelligence. We call the sum frequency the *upper* sideband and the difference frequency the *lower* sideband. The modulating signal in nearly all broadcasting equipment is a very complex waveshape. We generally indicate it as a sine wave of one fixed frequency, but actually it is constantly changing in both frequency and amplitude. If we use the fixed sine wave concept, we can illustrate the sidebands as in Fig. 12-6.

In this example, we have used a 5-kHz audio sine wave to

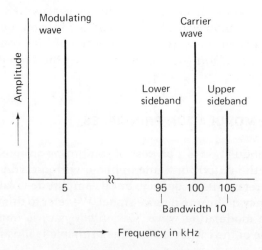

FIGURE 12-6. Sidebands

modulate a 100-kHz carrier wave. One hundred plus five gives us 105 kHz as the upper sideband frequency. One hundred minus five gives us 95 kHz as the lower sideband frequency. The difference between the two sideband frequencies is 10 kHz, and we call this the bandwidth. Since the modulating wave is nearly always a complex signal, these frequencies, except for the carrier wave, are changing all the time. However, if we think of them as an average, they give a reasonable facsimile of the real situation.

Degree of Modulation The degree of modulation is the amplitude ratio of the modulating wave voltage to the carrier wave voltage. We may express this ratio as an equation:

$$M = \frac{V1}{V2}$$

where M = degree of modulation
$V1$ = amplitude of modulating wave
$V2$ = amplitude of carrier wave.

We usually find the degree of modulation expressed as a percentage. The modulation percentage is simply the degree of modulation multiplied by 100. So 0.5 degree = 50 percent and 1 degree = 100 percent. When our carrier wave and modulating wave are the same amplitude, we have 1 degree or 100 percent modulation.

When we are checking the degree of modulation, we are generally observing a modulated carrier wave on an oscilloscope. The picture will resemble the drawing in Fig. 12–7.

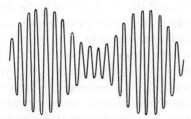

FIGURE 12–7. Modulated Carrier Wave

Since it is a simple matter to look at this modulated wave and determine the maximum and minimum voltage values, we need to restate our degree equation as follows:

$$M = \frac{V_{max} - V_{min}}{V_{max} + V_{min}}$$

With a minimum amplitude of 10 V and a maximum amplitude of 100 V, our degree of modulation is

$$M = \frac{V_{max} - V_{min}}{V_{max} + V_{min}}$$

$$= \frac{100 \text{ V} - 10 \text{ V}}{100 \text{ V} + 10 \text{ V}}$$

$$= \frac{90}{110}$$

$$= 0.818$$

Multiplying this result by 100 gives us 81.8 percent of modulation.

The ideal modulation is 1 degree or 100 percent. In order to accomplish it, the modulating wave must be the same amplitude as the carrier wave. If the carrier wave is larger, the modulation is *less than* 100 percent; if the modulating wave is larger, the modulation is *more than* 100 percent. With 100 percent modulation, our peak amplitude is exactly *twice* the amplitude of our unmodulated carrier wave. The minimum amplitude, at 100 percent modulation, goes to zero but only for an instant. Figure 12-8 illustrates over, under, and correct modulation.

AMPLIFIER MODULATORS

One technique used to accomplish amplitude modulation is to feed both the carrier wave and the modulating wave into a stage of radio frequency amplification. The carrier wave serves as our input signal, and we use the modulating wave to vary the bias on the amplifier. The result is an output carrier wave with varying amplitude. The variations in the carrier wave amplitude have the same frequency as the modulating wave. We can control the gain of an amplifier by injecting the modulating signal into the base, emitter, or collector of the amplifier transistor. Effectively, we are impressing our modulating wave upon the dc bias of the amplifier in such a way that any variation in the modulating amplitude causes a like change in the gain of the amplifier.

Base Injection When we inject the modulating signal into the base of a transistor amplifier, we are using the base injection method of modulation. An amplifier modulator of this type is illustrated in Fig. 12-9.

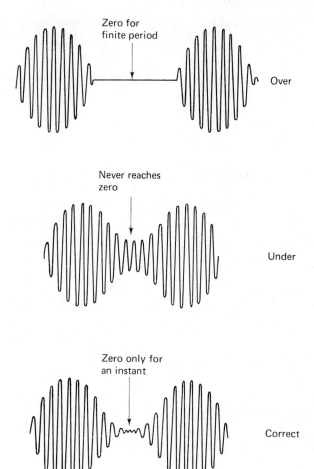

FIGURE 12-8. Modulation Waveforms

This is a common emitter, radio frequency (RF) amplifier. R1 and R2 form a voltage divider to provide proper input bias. R3 is a swamping resistor for stabilization, and C2 bypasses R3 to prevent degeneration. C3 and the primary winding of T2 form a tuned tank that serves as a collector load impedance. We have two inputs and one output. Input A is our carrier wave from an RF oscillator. The carrier wave is of a constant frequency and constant amplitude. This carrier wave is coupled across T1 to the base of our transistor. Input B is our modulating signal. The modulating signal is coupled through the secondary winding of T1 directly to the base of our transistor. C1 is an RF decoupling capacitor to provide an ac ground for the carrier frequency.

290 Modulators

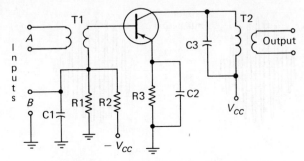

FIGURE 12-9. Base Injection Modulator

The carrier wave is being amplified, but the gain of our amplifier is varying. The modulating signal is alternately adding to and subtracting from the fixed bias. Since our modulating wave is generally an audio signal, we say that the gain of the amplifier is varied at an audio rate. The resultant waveform in our tank circuit is the carrier wave whose amplitude is varying at an audio rate. This modulated carrier wave is coupled from the secondary winding of T2 to the next amplifier stage.

Base injection is a method of low-level modulation. If we are to avoid distortion, the amplitude of the carrier wave must be less than the level of the dc bias between base and emitter. This means that the amplitude of our input waves must be very low. This is how we get the name of low-level modulation. The output from this modulator usually passes through several stages of amplification before it is broadcast from the antenna. It is difficult to obtain a high degree of modulation with the base injection method.

Emitter Injection Emitter injection is another low-level method of modulation. The actual circuit differs very little from that of the base injection modulator. The only real difference is the modulating signal is coupled to the emitter instead of to the base of the transistor. Such a modulator is illustrated in Fig. 12-10.

Compare the schematics in Fig. 12-9 and 12-10. Notice that all components are identical and carry the same identifying numbers in both schematics. Like-numbered components perform the same task in both circuits. Then where is the difference? Notice that input A, our carrier wave is still coupled across T1 to the base of our transistor. But input B, our modulating wave, is coupled directly to the emitter. The action is the same as previously described. The modulating wave varies the bias and causes the gain of the amplifier to vary at an audio rate. These variations are impressed into the amplitude of the carrier wave.

Amplifier Modulators   291

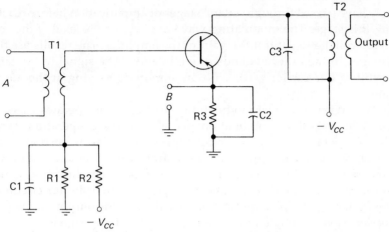

FIGURE 12-10. Emitter Injection Modulator

Collector Injection With a collector injection modulator, we can obtain either low-level or high-level modulation. Low-level modulation still deals with small signal amplitudes and then amplifies the modulated carrier. High-level performs modulation with high amplitude signals at a later point in the RF amplification chain. This type of modulator is illustrated in Fig. 12-11.

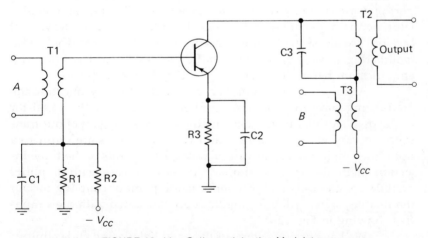

FIGURE 12-11. Collector Injection Modulator

Again, you will notice only a very slight change in the circuit. The only new component is T3, which we use for input coupling of input B, our modulating signal. The modulating signal now must match the amplitude of the carrier wave after amplification rather than before amplification. In fact, both carrier and modulating waves may have passed through several stages of amplification before reaching this stage. The modulating wave now varies the level of the collector voltage to control the gain of the amplifier. The results are the same as in the other methods of modulation. The signal in the tank circuit is our carrier wave with an amplitude varying at the audio rate.

With low-level modulation, our transistor operates class A, but with high-level modulation, we go to class C operation. With class C operation, the tank circuit supplies the missing portion of the carrier wave. With high-level modulation, we can attain a high degree of modulation, and in many cases, further amplification is not necessary. The modulation may take place in the collector circuit of the final stage of RF amplification. In this case, the output modulated carrier wave is coupled to the antenna as a driving signal.

OSCILLATOR MODULATORS

We can also amplitude modulate a carrier wave while it is being generated. The stage that accomplishes this operation is a combination oscillator and modulator which we call an oscillator modulator. The RF oscillator portion generates the carrier wave while the modulator portion varies the amplitude of the carrier wave. The oscillator modulator normally generates a high-level modulated carrier wave that can be broadcast with little additional amplification. The audio modulating wave is generally amplified through two stages of audio amplification before it is applied to the oscillator modulator.

Again we may accomplish our modulation by injecting the modulating wave into the base, emitter, or collector of the transistor in the oscillator modulator. Each method causes the gain of our transistor to vary at an audio rate, and these variations are impressed upon the amplitude of our carrier wave. The circuits of the three injection methods are very similar. As was the case with amplifier modulators, the only real difference is the point of coupling for our modulating signal. All three methods are illustrated with the simplified drawing in Fig. 12-12.

We have omitted some of the essential biasing and stabilization circuits, and, of course, we shall have only one input point for a particular oscillator modulator. C1 and the primary winding of T4 form our oscillating tank circuit. This tank is tuned to the desired carrier wave frequency. A tickler coil pickup loop on T4 provides the necessary positive feedback to sustain oscillations. The transistor provides the amplification. Without our modulating wave input, the

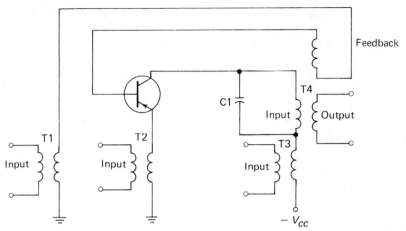

FIGURE 12-12. Methods of Injection

output would be a steady carrier wave of constant amplitude and frequency.

If we apply our modulating wave to T1, we are injecting the signal through the tickler coil to the base of the transistor. If we input the modulating wave through T2, we are injecting the signal to the emitter. If we send the modulating signal input through T3, we have collector injection. Either input will control the gain of the transistor and vary the amplitude of the carrier wave that is being generated. The output will be an RF carrier wave with its amplitude varying at the rate of the modulating wave.

Base Injection Since the three methods of injection have very similar circuits, a detailed explanation of the base injection system should suffice for all three. This system is illustrated in Fig. 12-13.

Our modulating wave from a microphone or some similar device has been amplified through two audio amplifier stages before being applied to the primary of T1. From the secondary of T1, the modulating wave is applied through the feedback coil of T2 to the base of our transistor.

The primary winding of T2, along with C3, is the oscillating tank circuit. This tank circuit determines the frequency of oscillation, which is our carrier wave frequency. The tank circuit is also the collector impedance for the transistor. The tickler coil picks up a portion of the carrier frequency signal and couples it back to the transistor base as regenerative feedback. Capacitor C1 is a very high impedance for the audio frequency of the modulating wave, and at the same time, it is a very low impedance decoupling for the carrier wave

294 Modulators

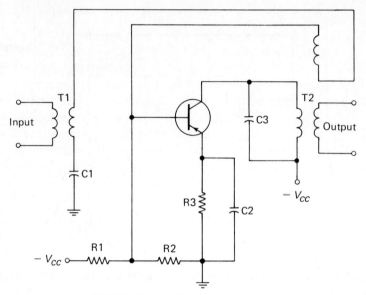

FIGURE 12-13. Oscillator Modulator

frequency. Our output from the secondary is an amplitude modulated carrier wave; the amplitude is varying at the audio rate of the modulating wave.

Resistors R1 and R2 form a voltage divider to provide proper input bias across the emitter-base junction. Stabilization is accomplished by swamping resistor R3. Capacitor C2 is an audio bypass capacitor to prevent degeneration of our modulating signal.

If we consider the oscillator in the absence of an input signal, we have an Armstrong RF oscillator. The transistor is alternately driven from cutoff to saturation and from saturation to cutoff. Feedback through the tickler coil is regenerative to sustain these oscillations. The output under these circumstances is a continuous carrier wave. It has a fixed frequency and a fixed amplitude.

Now when we apply our audio modulating wave through T1, it will vary the bias (and the gain) of the transistor. As the input swings negative, it increases the forward bias and causes the gain of the transistor to increase. During this negative alternation of the modulating signal, the carrier wave oscillations increase in amplitude. As the input swings positive, it decreases the foward bias and reduces the gain of the transistor. During this positive alternation of the modulating signal, the carrier wave oscillations decrease in amplitude. Thus, the amplitude of the carrier wave being generated is caused to vary at the same rate as our modulating wave.

When our gain increases, it causes the collector voltage to drop to a lower level. When gain decreases, our collector voltage increases. This shift in collector voltage causes a change in the collector-emitter capacitance in the transistor. The capacitance decreases when gain decreases, and vice versa. When the capacitance decreases, it causes an increase in the resonant frequency of the tank circuit. When the capacitance increases, it causes a decrease in the resonant frequency of the tank circuit. Since our resonant frequency is changing, we must also be changing the frequency of the carrier wave. This amounts to modulating our frequency at the same time that we modulate the amplitude. This frequency modulation is an undesirable feature of the oscillator modulator, but the degree of frequency modulation is low and can be tolerated.

FREQUENCY MODULATION PRINCIPLES

When we frequency modulate a carrier wave, we hold its amplitude constant and cause the frequency to vary at an audio rate. This operation transfers the intelligence of the modulating wave into the varying frequency of the carrier wave. One advantage of frequency modulation (FM) over AM is the relatively static-free transmissions.

The Carrier Wave Remember that the oscillator modulators that we discussed for AM produced both amplitude and frequency modulation. A similar oscillator produces and modulates our FM carrier wave. This means that we have some undesirable amplitude changes. However, the amplitude modulation can be easily removed by a clipping (limiting) action. An oscillator modulator is generally used to frequency modulate the carrier wave as it is being generated. The amplitude modulation is then removed by clipping. The carrier wave then is of a constant amplitude but varying frequency. This FM carrier wave is a relatively weak signal that must be amplified through several stages. In most cases, the carrier wave output from the modulator is at a frequency much too low for transmission. In this case, we send it through frequency multiplier stages to increase the frequency. The steps in processing the carrier wave are illustrated in Fig. 12-14.

Waveshape A is an audio wave from our microphone that is coupled into the oscillator modulator. The carrier wave being generated will be altered in both frequency and amplitude at the same rate as the frequency of our modulating wave. The result is similar to waveshape B. The changes in amplitude of this wave are undesirable,

296 Modulators

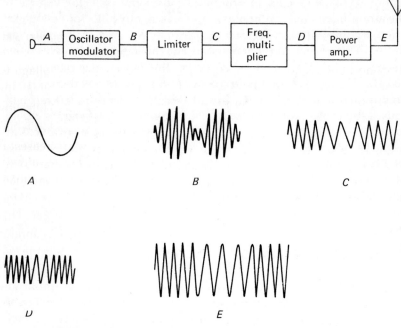

FIGURE 12-14. Generating and Processing an FM Carrier Wave

so we send it through a limiter. The limiting action removes all peaks that exceed a preset value and produces waveshape C. Now we have a carrier wave of a constant amplitude but varying frequency. The frequency multiplier takes waveshape C and boosts the frequency to the desired transmission frequency. This is waveshape D, a high-frequency, constant amplitude, frequency modulated carrier wave. This is the carrier wave that we wish to transmit, but it has insufficient power. The power amplifier boosts the wave in amplitude and supplies the necessary power for transmission. The result is waveshape E, which is coupled to the antenna.

Frequency Deviation We refer to the unmodulated carrier wave frequency as the *center* frequency or the *rest* frequency. This is the same as the resonant frequency of the tank circuit in the oscillator modulator when there is no input signal. When we bring in the modulating wave, it causes the carrier wave frequency to deviate to either side of the center frequency. The number of times per second that the frequency goes through a complete deviation cycle is called the *rate* of deviation. This rate of deviation is determined by the frequency of our modulating wave. The number of hertz between center and minimum or maximum frequency of the carrier wave is known as the *degree* of deviation. The degree of deviation is controlled by the amplitude of our modulating signal. Both rate and degree of deviation are illustrated in Fig. 12-15.

This drawing represents a carrier wave with a center frequency of 100 MHz that is being modulated by a single tone of 500 Hz. The 500-Hz modulating wave is represented at two different amplitudes. Consider waveshape A. On its positive alternation, it causes the carrier wave frequency to increase from 100 MHz (center) to 100.025 MHz and decrease back to the center frequency. On its negative alternation, it causes the carrier frequency to decrease to 99.975 MHz and increase back to the center frequency. The degree of deviation is 25 kHz; our carrier wave frequency is varying 25 kHz to either side of the center frequency. The rate of deviation is 500 Hz; our carrier wave goes through a complete deviation cycle 500 times per second. Notice that the rate of deviation is the same as the frequency of our modulating signal.

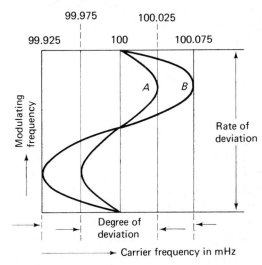

FIGURE 12-15. Frequency Deviation

Now consider waveshape B. It is the same frequency but twice the amplitude of waveshape A. On the positive alternation it causes the carrier frequency to increase from 100 MHz (center) to 100.075 MHz and decrease back to the center frequency. On the negative alternation it causes the carrier wave frequency to decrease to 99.925 MHz and increase back to the center frequency. Our degree of deviation is now 75 kHz, and our rate of deviation is still 500 Hz, because we still have the same modulating frequency.

Remember in this example our modulating wave was a

single tone of 500 Hz. If we are broadcasting intelligence, especially speech or music, the modulating wave will be a complex audio wave that constantly changes in both amplitude and frequency. This means that both rate and degree of deviation are constantly changing.

Sidebands In AM we have two significant sidebands; one above and one below the center frequency of the carrier wave. In FM we generate many sidebands both above and below the center frequency of the carrier wave. The center frequency of the carrier wave has no intelligence; all the intelligence is carried in the sidebands. This makes all the sidebands significant until the amplitude decreases to almost zero. We say that any sideband with an amplitude of one percent of the carrier wave center frequency is a significant sideband. A center frequency and a number of sidebands are illustrated in Fig. 12–16.

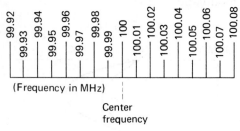

FIGURE 12–16. FM Sidebands

Here we have a carrier wave with a center frequency of 100 MHz, and we are modulating it with a single tone of 10 kHz. On the high side of the center frequency, we have a separate sideband for each multiple of the modulating wave frequency. The first sideband is 100 MHz + 10 kHz, or 100.01 MHz. The second is 100 MHz + (2 × 10 kHz), or 100.02 MHz. The eighth sideband is 100 MHz + (8 × 10 kHz), or 100.08 MHz. On the low side of the center frequency, we have a separate sideband for each submultiple of our modulating frequency. The first low sideband is 100 MHz − 10 kHz, or 99.99 MHz. The second sideband is 100 MHz − (2 × 10 kHz), or 99.98 MHz. The eighth sideband is 100 MHz − (8 × 10 kHz), or 99.92 MHz.

Theoretically, these sidebands extend in both directions indefinitely, and there is no definite pattern as to where they are strong and where they are weak. However, as we move outward from the center frequency, the power of the sidebands does grow weaker. When the amplitude of the sidebands drop to one percent of that of the center frequency, they no longer have enough power to make them significant.

Bandwidth Ideally, we should have a bandwidth to include all significant sidebands. However, in FM broadcasting the bandwidth is a legal rather than a convenient consideration. In the United States, the Federal Communications Commission has fixed the maximum bandwidth at 150 kHz. When we apply this legal bandwidth to the example of Fig. 12–16, we see that it includes seven pairs of sidebands (seven upper sidebands and seven lower sidebands). The eighth pair of sidebands is outside the legal bandwidth and must be eliminated. This may or may not include all the significant sidebands. Any significant sideband that is excluded means a loss of some quantity of intelligence. This loss of intelligence will show up as distortion in the received signal.

The percentage of modulation is also related to the legal bandwidth. When we have a deviation of 75 kHz to either side of the center frequency, we have reached our legal limit, a 150-kHz bandwidth. We consider this to be 100 percent modulation. It may not be the best that we can get, but it is the best that we are allowed to use. A bandwidth of 75 kHz is considered 50 percent modulation.

Modulation Index We define modulation index as the ratio of deviation to the highest modulating frequency.

$$M = \frac{fd}{fm}$$

where M = modulation index

fd = frequency deviation of carrier

fm = maximum modulating frequency.

In the case represented in Fig. 12–15, waveshape B is a modulating wave of 500 Hz, and it causes a frequency deviation of 75 kHz. The modulation index is

$$M = \frac{fd}{fm}$$
$$= \frac{75 \text{ kHz}}{500 \text{ Hz}}$$
$$= 150$$

This modulation index is a reference number. It refers to a table that gives us the number of significant sidebands and the bandwidth necessary to include all these sidebands. We call this table a modulation index table, and a sample of it is represented in Fig. 12–17.

300 Modulators

Mod. index	Sig. sidebands	Bandwidth (F = mod. freq.)
0.5	4	4 × F
1	6	6 × F
2	8	8 × F
3	12	12 × F
4	14	14 × F
5	16	16 × F
6	18	18 × F
7	22	22 × F

FIGURE 12-17. Modulation Index Table

The modulation index, as previously calculated, is located in the left column of the table. We locate the modulation index in this column and then move across the page. The second column gives us the number of significant sidebands. The third column is simply the number of significant sidebands multiplied by the modulating frequency. If we have a modulation index of 6 and a modulating frequency of 5 kHz, we have 18 significant sidebands and require a bandwidth of 18 × 5 kHz. This bandwidth of 90 kHz is well within the 150-kHz legal limit.

FREQUENCY MODULATORS

All of the oscillator modulators that we discussed for AM can be used to generate both amplitude and frequency modulation. However, they were designed for AM, and the FM is an undesirable side effect. An oscillator modulator that is designed for FM will have some AM, as previously discussed, but it does a better job of frequency modulation.

Capacitor Microphone Modulator The capacitor microphone modulator is an oscillator modulator that illustrates the principles of frequency modulation. It has no other useful function, since it is not a practical circuit. However, it should be instructive for us to analyze the capacitor microphone modulator at this time. The circuit is illustrated in Fig. 12-18.

The capacitor microphone is a variable capacitor. It has one fixed plate and one movable plate. Sound waves striking the diaphragm of the microphone causes it to compress and release. This

Frequency Modulators 301

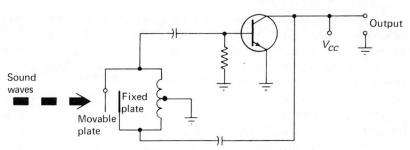

FIGURE 12-18. Capacitor Microphone Modulator

varies the spacing between the plates and causes the capacitance to vary in accordance with the spacing. When a compression wave strikes the diaphragm, it presses the plates closer together. The reduced spacing increases the capacitance and decreases the resonant frequency of the tank circuit. When a rarefaction wave is present, it allows the movable plate to move outward and increases the spacing between the plates. This increase in spacing causes a decrease in capacitance and an increase in the resonant frequency of the tank circuit.

The tank circuit is a frequency determining device in a Hartley oscillator. The oscillator has no choice but to oscillate at the ever-changing frequency of the tank circuit. The carrier wave generated by the oscillator has a frequency variation that incorporates every aspect of the amplitude and frequency of the modulating wave.

Reactance Modulator The reactance modulator provides a more practical method of frequency modulating a carrier wave. This circuit incorporates a variable reactance transistor and an Armstrong oscillator. A simplified schematic is illustrated in Fig. 12-19.

We have omitted the normal biasing and stabilizing circuits

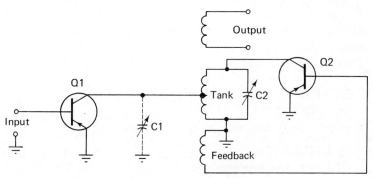

FIGURE 12-19. Reactance Modulator

better to illustrate the action. The oscillator portion of the circuit consists of Q2, a tank circuit, and a feedback loop. Capacitor C2 is tuned for the desired center frequency of the carrier wave without the input modulating wave. The oscillator, without the modulating input, is a simple Armstrong oscillator.

Transistor Q1 is our variable reactance component. Capacitor C1 represents the collector to emitter capacitance. This capacitance varies with the bias, and the bias varies in accordance with the modulating wave input. Notice that C1 parallels about half of the turns on the coil of the tank circuit. Therefore, any change in C1 alters the resonate frequency of the tank circuit. As this resonate frequency changes, the frequency of the carrier wave deviates from the center frequency. The coil of the tank circuit is the primary winding of a transformer. The feedback loop and the output coil are secondary windings of the same transformer.

REVIEW EXERCISES

1. Differentiate between modulation and modulator.
2. Name the two principal types of modulation.
3. How is a microphone related to modulation?
4. Describe the function of a carbon microphone.
5. What causes the background hiss in a carbon microphone?
6. Why is the carbon microphone acceptable for voice but undesirable for music?
7. Why do carbon microphones require a transformer coupling when feeding into an amplifier?
8. The crystal microphone is based on what principle?
9. Why is the coupling problem between a crystal microphone and an amplifier simpler than that between a carbon microphone and an amplifier?
10. What is the principal disadvantage of the crystal microphone?
11. What is the principle of the dynamic microphone?
12. What is a carrier wave?
13. What is a modulating wave?
14. Differentiate between amplitude modulation and frequency modulation.
15. In amplitude modulation, what constitutes the
    a. Upper sideband?
    b. Lower sideband?
16. How do we determine the bandwidth in amplitude modulation?

17. A 200-kHz carrier wave is amplitude modulated by a 10-kHz intelligence wave. What is the frequency of the
    a. Upper sideband?
    b. Lower sideband?
18. What is the bandwidth in the case of item 17?
19. What is meant by degree of modulation?
20. Our carrier wave has an amplitude of 50 V and our modulating wave has an amplitude of 45 V. What is the
    a. Degree of modulation?
    b. Percentage of modulation?
21. An amplitude modulated carrier wave on an oscilloscope is varying in amplitude from 5 V to 100 V. What is the
    a. Degree of modulation?
    b. Percentage of modulation?
22. Observe the waveshapes in Fig. 12–20 (page 304) and identify the wave which represents
    a. Overmodulation.
    b. Undermodulation.
    c. Correct modulation.
23. Describe the process of amplitude modulation in an amplifier.
24. Name the three methods of modulation in amplifier modulators.
25. Of the three methods, which are low-level and which are high-level?
26. Explain the difference between low-level and high-level modulation.
27. Describe the function of an oscillator modulator.
28. To a certain degree, we get both AM and FM when we modulate a carrier for either AM or FM. What do we do about
    a. FM with amplitude modulation?
    b. AM with frequency modulation?
29. What is frequency deviation?
30. Differentiate between rate of deviation and degree of deviation.
31. What determines the deviation
    a. Rate?
    b. Degree?
32. A 200-MHz carrier wave is frequency modulated by a 10-kHz tone. What is the frequency of each of the
    a. First pair of sidebands?
    b. Second pair of sidebands?

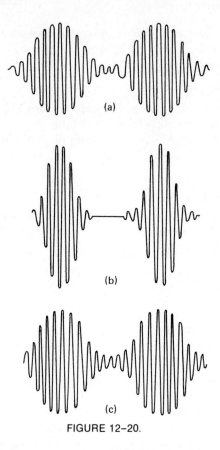

FIGURE 12-20.

33. When do we consider FM sidebands to be insignificant?
34. What determines the maximum bandwidth that can be broadcast when one is using FM in the United States?
35. Define modulation index.
36. A carrier wave is frequency modulated by a 10-kHz tone, and the carrier frequency is swinging 50 kHz to either side of center. What is the modulation index?
37. How do we determine the bandwidth and number of significant sidebands after calculating the modulation index?
38. Describe the action of a variable reactance modulator.

Chapter Thirteen

Demodulators

In the last chapter, we discussed methods of modulating a carrier wave. That carrier wave, after proper processing, is applied to an antenna and radiated into space. At the receiver, the carrier wave cuts across the receiving antenna and generates a current into the antenna. This generated current is usually very weak, but it has all the characteristics of the transmitted carrier wave. Now the task is to process this weak signal, extract the intelligence, and reproduce the intelligence that formed the modulating wave in the transmitter. The process of extracting the intelligence from a modulated carrier wave is called demodulation. The device that is used for this extraction process is a demodulator.

The intent of this chapter is to analyze demodulation circuits and principles. Along the way, we shall examine some of the other features of processing a received signal. Since we had different types of modulation in transmitters, we must have different types of demodulators in receivers. The AM receiver is designed to receive and process only AM signals. The FM receiver is designed to receive and process only FM signals. We shall consider AM and FM demodulation separately.

RECEIVER FUNCTIONS

Whether AM or FM, a receiver performs four basic functions: reception, selection, demodulation, and reproduction. These four functions are illustrated in Fig. 13-1.

Most of our receiver antennas are not very selective. Any and all electromagnetic waves that drift by will induce a current into the antenna. In fact, current is induced into any conductor that an

306 Demodulators

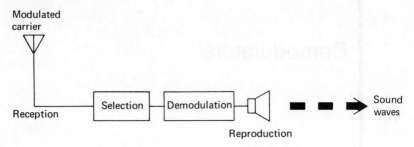

FIGURE 13-1. Receiver Functions

electromagnetic wave cuts across. Reception is accomplished by placing a receiver antenna in a position to encounter the electromagnetic wave.

Our receiver must select the desired carrier wave from among the many signals that are continuously cutting across the antenna. This is the job of the receiver's selection circuits. A receiver's selectivity rating is an indication of its ability to differentiate between desired and undesired signals.

Demodulation is the process of removing the low-frequency intelligence from the high-frequency carrier. This demodulation section is where AM receivers are vastly different from FM receivers. The result, in either case, is an audio frequency signal equivalent to the modulating signal being used in the transmitter.

The audio frequency signal is coupled into a device that converts electrical energy to sound energy. These devices are microphones in reverse; we call them speakers. They may be large speakers mounted in separate cabinets or very small speakers mounted in headphones. The results are the same; they take the audio frequency signal and produce the sound equivalent of this electrical signal. The sounds coming from the speaker should be exact reproductions of the sounds that are striking the microphone in the transmitter.

Heterodyning Most of our commercial broadcast band receivers that are designed for AM reception perform the task of selection through a process that we call heterodyning. We refer to receivers of this type as superheterodyne receivers. Heterodyning is a process of mixing two frequencies together in order to produce a third frequency. A signal of one frequency is generated locally in an oscillator circuit. The second signal is the carrier wave from the antenna. The third signal comes from the mixing of these two, and it is the difference between the first two. This difference frequency falls within the selected bandwidth only when the received signal is of the exact frequency that we are trying to tune in.

The mixing process is very similar to the action in a modulator. A great variety of frequencies are generated, but the tuned filter circuits remove everything that falls outside of a narrow bandwidth. This bandwidth should be wide enough to include the difference frequency with both upper and lower sidebands. The heterodyning action is illustrated in Fig. 13-2.

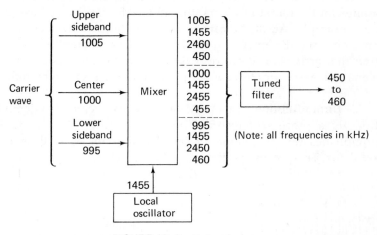

FIGURE 13-2. Heterodyning

We have assumed a carrier wave of 1000 kHz modulated with a single 5-kHz tone. This gives us sidebands of 995 kHz and 1005 kHz. The local oscillator is producing a continuous signal of fixed amplitude at a frequency of 1455 kHz. The local oscillator signal is mixed with all components of the carrier wave frequency. When mixed with the upper sideband frequency (1005 kHz), it produces a sum of 2460 kHz and a difference of 450 kHz. In addition, we still have the two original frequencies. When the 1455 kHz mixes with the carrier wave center frequency (1000 kHz), it produces a sum of 2455 kHz and a difference of 455 kHz. In addition, we still have the two original frequencies. When the 1455 kHz mixes with the lower sideband frequency (995 kHz), it produces a sum of 2450 kHz and a difference of 460 kHz. In addition, we still have the two original frequencies.

Of course, many other frequencies are generated, but these will suffice for illustration. The output of the mixer stage is a tuned filter with a band width just wide enough to accommodate the band-

308 Demodulators

width of the received signal. In this case, we have a 10-kHz bandwidth, and the tuned filter passes only the frequencies from 450 to 460 kHz. Notice that this bandwidth includes the difference frequency from the carrier center frequency and from both sidebands. All other frequencies are filtered out.

We have now eliminated all of the radio frequencies and reduced our signal to a relative low frequency that we call the intermediate frequency. When we refer to the intermediate frequency (IF), we are generally speaking of the center of the IF band. For most commercial broadcast receivers, this IF is 455 kHz, as we have used in our example. All of the carrier wave intelligence has been transferred onto the IF signal, and the IF has the same amplitude modulation that arrives on the carrier wave. This IF signal goes through several stages of amplification before it reaches the demodulator.

Demodulation In an AM receiver, we accomplish demodulation by a process of rectification and filtering. The circuit that accomplishes this is commonly called a detector as well as a demodulator. The process is illustrated in Fig. 13-3.

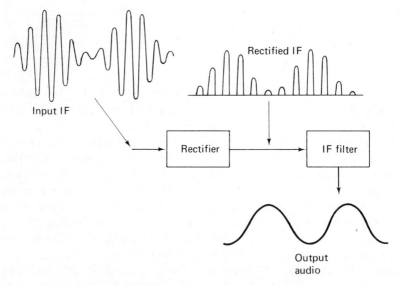

FIGURE 13-3. Demodulation

Our input signal from the final IF amplifier is an IF signal (center frequency of 455 kHz) that is amplitude modulated with our original audio signal. The rectifier portion of the demodulator clips off the negative portion of each IF cycle. This leaves only the positive half-cycles, but the audio variations are still apparent on the positive peaks. The filter circuit removes all components of the IF signal and leaves only the audio envelope. This audio signal should be an exact reproduction of the modulating signal being used in the transmitter. The speaker converts this audio signal back into sound waves.

MIXERS AND CONVERTERS

When heterodyning is accomplished by using two separate stages as illustrated in Fig. 13-2, we call these stages local oscillator and mixer. The mixing takes place in the mixer stage. This is similar to the modulation process, where we use two inputs to an amplifier modulator. But heterodyning can be accomplished in only one stage. When this is done, we call the stage a *converter*. The converter combines the functions of both local oscillator and mixer into a single stage. The converter action is similar to modulation in an oscillator modulator.

Base Injection Mixer We stated that a mixer stage has two inputs: one from the local oscillator and one from the antenna. The input from the local oscillator is a continuous RF signal. When we tune the mixer to accept a certain frequency, the same control tunes the local oscillator to produce a signal 455 kHz (the IF) above the received frequency. The input from the antenna is generally amplified through one or more preamplifiers before it is applied to the mixer. We may inject these signals to any element of the transistor. Figure 13-4 illustrates base injection for both the inputs.

When we turn the tuning control on our receiver to select a particular station, we are varying the capacitance of C1. C1 and the primary winding of T1 form a tuned parallel tank circuit that accepts only the frequency for which it is tuned. (Actually, we are describing an ideal condition. In practice, we pick up stray signals and some slightly off-frequency stations.) Capacitor C1 is ganged to a similar capacitor in the frequency determining tank circuit of the local oscillator. When we tune C1 for a specific frequency, we also tune the local oscillator for a frequency 455 kHz above the selected carrier wave frequency. Thus, the frequency from our local oscillator is always 455 kHz above the center frequency of the carrier wave from the antenna.

The modulated carrier wave is coupled across T1, while the steady RF from the local oscillator is coupled across T2. Both of these signals are applied directly to the base of our transistor. The tank

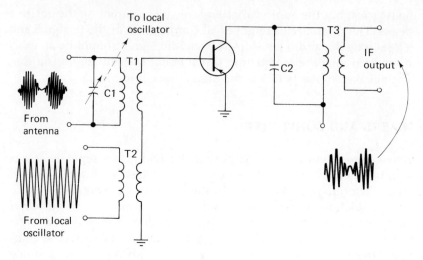

FIGURE 13-4. Simplified Base Injection Mixer

circuit in the collector of the transistor is tuned to 455 kHz with a bandpass from 450 to 460 kHz. This represents the difference frequency between the two inputs. This difference frequency with a center frequency of 455 kHz will be amplified through the transistor and coupled across T3 as an input to the IF amplifiers. The IF output has the same amplitude modulation that we impressed on the carrier wave in the transmitter. The intelligence has been transferred from the RF carrier wave to the IF wave. We shall now amplify the IF signal through several stages before we remove the intelligence with a demodulator.

This simplified mixer circuit *is not* desirable for two reasons: First, the carrier wave couples back to the local oscillator and interferes with its frequency; second, the local oscillator frequency couples back to the antenna and radiates an interference signal.

Modified Base Injection Mixer The modified base injection mixer has most of the shortcomings just described, but we shall discuss it briefly. It is illustrated in the schematic of Fig. 13-5.

In this schematic, we have included the necessary biasing and stabilizing circuits. R1 and R2 form a voltage divider for input bias, and C3 provides an RF ground. R3 is a swamping resistor for stabilization, and C4 bypasses this resistor to prevent degeneration.

The input from the local oscillator is coupled across C1 to the base of the transistor. The input from the antenna is coupled across T1 and is also applied to the base. C1 offers greater opposition to the

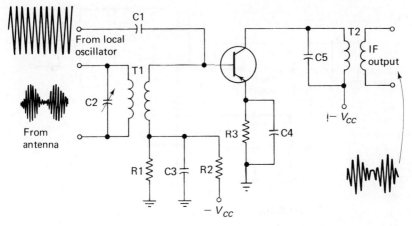

FIGURE 13-5. Modified Base Injection Mixer

modulated carrier wave than it does to the local oscillator signal, and to some extent isolates the local oscillator from interference. Some of the local oscillator signal may still couple across T1, secondary to primary, and radiate interference from the antenna.

The mixing action takes place as previously described, and the collector tank responds to the difference frequency. The output is the same IF with a center frequency of 455 kHz, and its amplitude is varying at our modulating rate.

Emitter Injection Mixer The best way to stop the modulated carrier wave from interfering with the local oscillator frequency is to input the signals on different elements of the transistor. At the same time, this arrangement can be used to isolate the local oscillator from the antenna and prevent the radiation of interference signals. Both of these desirable features are incorporated into the emitter injection mixer, as illustrated in Fig. 13-6.

Here we have the modulated carrier wave from the antenna transformer coupled to the base of our transistor. The continuous RF signal from the local oscillator is RC coupled to the emitter of our transistor. This arrangement effectively isolates the local oscillator from the incoming signal, and isolates the antenna from the local oscillator signal. Otherwise the circuit is the same as those previously discussed. The mixing is the same, and the output is still an amplitude modulated IF signal with a center frequency of 455 kHz.

Practical Mixer Figure 13-7 is a complete schematic of a more practical mixer circuit than any we have previously discussed.

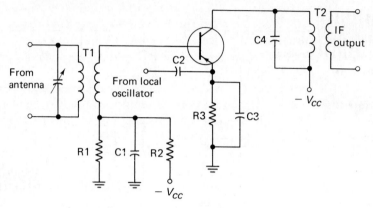

FIGURE 13-6. Emitter Injection Mixer

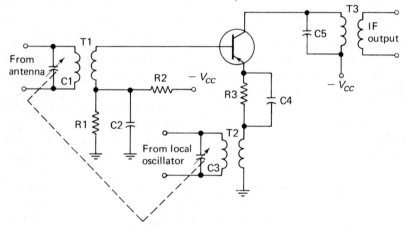

FIGURE 13-7. Practical Mixer

First notice that C1 and C3 are ganged together by a mechanical connection; when we tune one, we tune both. This same mechanical connection tunes the frequency of the local oscillator. In fact, the tank circuit formed by C3 and the primary of T2 could be the collector tank circuit in the local oscillator.

The modulated carrier wave from the antenna is tuned in by varying C1. This wave is then coupled across T1 to the base of our transistor. When we tune in a station by varying C1, we also vary C3 so that the local oscillator frequency is 455 kHz above the incoming carrier wave frequency. This local oscillator signal is coupled across T2 to the emitter of our transistor. With this arrangement, we can mix the two signals without any type of interaction. The carrier wave is isolated from the local oscillator, and the local oscillator signal is isolated from the antenna.

The mixing action takes place as we previously discussed. The result is an IF signal with the intelligence impressed upon its amplitude. This IF signal is coupled across T3 and into the first IF amplifier. After several stages of IF amplification, we shall couple it into a demodulator and extract the intelligence.

The remainder of the circuit is standard biasing and stabilization components. R1 and R2 form a voltage divider for the input bias. C2 is an RF short to keep the RF out of the dc power supply; it provides an RF ground. R3 is a swamping resistor for stabilization, and C4 is a bypass capacitor to prevent degeneration.

Converters As previously stated, when we incorporate the local oscillator and mixer into a single stage, we call it a converter. The only input to a converter is the modulated carrier wave from the antenna. The local oscillator signal that mixes with the input is generated within the converter stage. A schematic for a practical converter is illustrated in Fig. 13–8.

The oscillating circuit composed of C4, C5, and the secondary winding of T2 takes the place of our local oscillator. The continuous RF from this tank circuit is coupled through C3 to the emitter of the

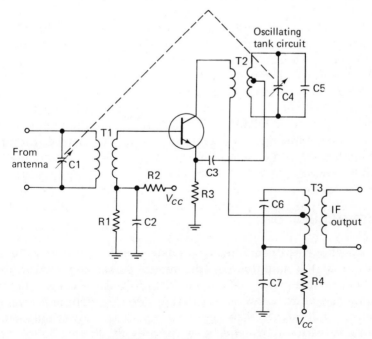

FIGURE 13–8. Practical Converter

transistor. The primary winding of T2 is a feedback coil. The RF energy coupled across T2 is sufficient to sustain the oscillations in the tank circuit. The collector load impedance is composed of R4 and the primary windings of T2 and T3. Direct current bias for the input is furnished by the voltage divider, which consists of R1 and R2. R3 is a stabilizing resistor, and since it is not bypassed, it does cause some degree of degeneration.

The input amplitude modulated carrier wave is applied to the primary winding of T1. This winding forms a tuned tank circuit with C1. C1 is variable, and it is ganged with one of the capacitors in the oscillating tank circuit. When we turn the selection control to tune in a particular station, we vary both of these capacitors. The frequency of the oscillating tank circuit tracks 455 kHz above the selected incoming carrier frequency.

The incoming carrier wave is coupled across T1 to the base of the transistor. The oscillating tank frequency is coupled into the emitter. The signals mix through the transistor, and many frequencies may be detected at the collector. One of these frequencies reinforces the oscillations by coupling across T2 into the tank circuit. The most prominent frequency is the 455 kHz that is the difference between the carrier wave frequency and the oscillating tank frequency.

The primary winding of T3 and C6 form another tuned tank circuit. This parallel tank is tuned to a 455-kHz center frequency with a bandpass from about 450 to 460 kHz. This tank circuit represents the principal impedance in the collector circuit. The 455 kHz will be emphasized over all other frequencies, and this is the only frequency that couples across T3. The output is an IF signal with a center frequency of 455 kHz. This IF signal is amplitude modulated with the same intelligence that arrives on the carrier wave from the transmitter. C2 and C7 are RF decoupling capacitors to protect the dc power supply from RF oscillations.

AM DEMODULATORS

The job of our AM demodulator (or detector) is to extract the intelligence from the amplitude modulated IF signal and recreate our original modulating signal. Demodulation consists of two distinct operations: rectification and filtering. Rectification is necessary because the IF signal is composed of upper and lower sidebands. The signal varies above and below zero by the same amount. This configuration gives us an average ac value of zero. The first step in extracting the intelligence is to clip off half of the IF waveshape. When we filter out the IF from the remaining half, we have an envelope of the amplitude variations. This envelope is an ac sine wave that is equivalent to the modulating signal being used in the transmitter.

Series Diode Demodulator The series diode demodulator uses a diode in series with a load resistor. We shall consider this circuit without a filter to clarify the rectifying function. Then we shall add a filter to remove the IF components. Figure 13-9 shows the circuit without a filter.

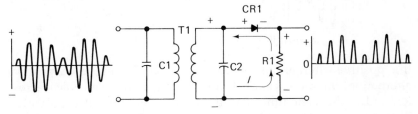

FIGURE 13-9. Demodulator Rectification

The amplitude modulated IF signal from the final stage of IF amplification is coupled across T1. Both the primary and secondary windings of T1 are tuned tank circuits, and they are tuned to the 455-kHz center frequency of the input. On each positive alternation of each IF sine wave, the indicated polarities are developed in the circuit. During these positive alternations, current will flow as indicated. The quantity of current through R1 is determined by the amplitude of the particular positive alternation. As a result of this action, each positive alternation of the input is developed across R1. During each negative alternation of the IF, the indicated polarities will reverse and place reverse bias on the diode. All current ceases during these negative alternations, and the output is zero during these times. The resulting output is a nearly perfect reproduction of all the positive alternations of the IF signal. All of our intelligence is still contained in the varying amplitude of the rectified waveshape. We now need a filter to remove the IF components from the rectified waveshape and to recover the audio modulating wave. We can complete the circuit by adding a capacitor in parallel with R1, as shown in Fig. 13-10.

The only difference between this circuit and that of Fig. 13-9 is the addition of C3 in parallel with R1. But notice that this one small change makes a tremendous difference in our output waveshape. The solid arrows on this diagram show the current flow on

316 Demodulators

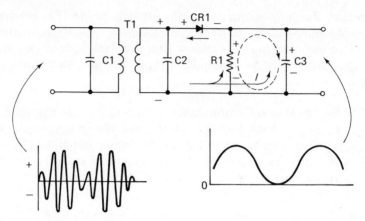

FIGURE 13-10. Rectifying and Filtering

the positive alternations of our IF signal. During each positive alternation, the diode conducts, developing a positive voltage across R1, and at the same time charging C3 with the indicated polarities. This charge on C3 in each instant will be equal to the peak value of that alternation.

During the negative alternations of the IF signal, the diode cuts off, and the capacitor tries to discharge through R1. This discharge path is indicated by the broken arrow. R1 and C3 form a long RC time constant for the IF signal. As a result, the capacitor has a chance to lose only a small portion of its charge between charging cycles. The action of C3 removes all the IF components and leaves only the audio envelope. This audio is an almost perfect reproduction of the modulating wave being used in the transmitter.

Shunt Diode Demodulator The shunt diode demodulator is constructed by placing a rectifying diode in parallel (shunt) with the load resistor. This type of demodulator normally uses a series inductor for a filter. This arrangement is illustrated in Fig. 13-11.

With a shunt diode used in this fashion, most of our output is developed when the diode is cut off. When our diode conducts,

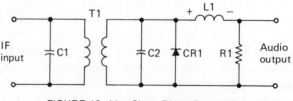

FIGURE 13-11. Shunt Diode Demodulator

the output is shorted out except for the filtering action of L1. Each positive alternation of the IF signal places reverse bias on the diode and cuts it off. During this time, the positive alternation is developed across R1, and the inductor L1 charges as indicated. On the negative alternations, the diode conducts, and except for L1, all current through the resistor would cease. However, the decaying field of L1 reverses its polarity and keeps current moving in the same direction. The output waveshape is about the same as that in Fig. 13-10. The IF components are filtered out, leaving only the audio envelope.

Transistor Demodulator A diode demodulator is *never* 100 percent efficient; the output amplitude is always slightly less than 50 percent of the input amplitude. When we have relatively large input signals, this slight loss is of little consequence. However, when we wish to demodulate a weak signal, the normal diode loss may be intolerable. In this case, it becomes necessary to add amplification to our demodulator stage. We can do so by using a transistor instead of a diode. The transistor is much more sensitive to small signals, and it increases the signal amplitude. Such a demodulator is illustrated in Fig. 13-12.

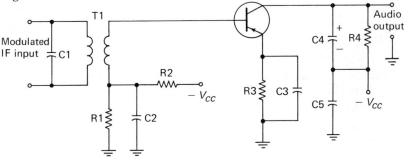

FIGURE 13-12. Transistor Demodulator

At first glance, this appears to be nothing more than an amplifier. It does amplify; that is one of its primary functions. But it is much more than just an amplifier. This stage has three functions: amplification, rectification, and filtration.

Let us examine the familiar circuit components. R1 and R2 form a voltage divider for input bias. This fixed bias is set so that the transistor is very nearly (but not quite) cut off. C1 is a bypass capacitor to protect the dc power supply from ac components; C5 serves this same function in the collector circuit. R3 is our stabilizing swamping resistor, and it is bypassed by C3 to prevent degeneration.

Our modulated IF signal is coupled across T1 to the base of

the transistor. Each negative alternation increases the forward bias, causes the transistor to increase conduction, and is amplified through the transistor. We have phase inversion through a common emitter, so each negative input pulse gives us a positive output pulse. Each pulse leaves capacitor C4 charged as indicated. Each positive alternation of the input places reverse bias on the transistor and cuts it off. This stops the collector current, and C4 starts to discharge through R4. This filtering action removes all IF components and leaves only the audio envelope. The input and output waveshapes are the same as those in Fig. 13–10. The principal advantage of this circuit is its ability to demodulate a very weak signal. Making it sensitive enough to respond to weak signals makes it incapable of handling large signals. A large signal would drive our transistor into saturation and cause distortion of the intelligence.

FM DEMODULATORS

Although we have some minor differences in the antenna and IF amplifiers, the main difference between an AM and an FM receiver is in the demodulator. You probably anticipated this from the fact that the AM intelligence is extracted from the carrier amplitude, whereas the FM intelligence is extracted from the carrier frequency. You will recall, from our chapter on modulators, that FM carrier waves vary in frequency according to the information being transmitted.

In the receiver, the FM signal is mixed with a local oscillator signal to produce the IF signal. This heterodyning is accomplished by either a mixer or a converter. During the heterodyning process, the frequency modulation of the carrier wave is transferred to the IF signal. We process the IF signal through several stages of IF amplification and then demodulate it in order to recover the intelligence.

The FM demodulator is often referred to as a *frequency discriminator*, because it reacts to a change in frequency. The function of an FM demodulator is to generate an audio sine wave with frequency proportional to the rate of frequency deviation and amplitude proportional to the degree of deviation.

Foster Seeley Discriminator (demodulator) The Foster Seeley discriminator is one of the most widely used FM demodulators. A schematic of this demodulator is shown in Fig. 13–13.

Basically the Foster Seeley discriminator converts the frequency changes into a changing voltage, rectifies the resulting ac, and filters out the IF components. The arrangement of the three coils

FM Demodulators   319

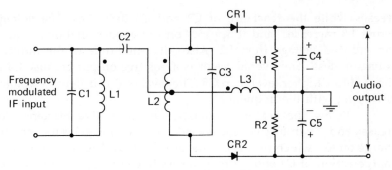

FIGURE 13-13.   Foster Seeley Discriminator

on a special transformer converts the frequency changes to voltage changes. These three coils are L1, L2, and L3. The tank circuit composed of C1 and L1 is the transformer primary, and it is tuned to the IF signal's center frequency. C3 and L2 form another tank circuit, and it is also tuned to the center frequency. The frequency modulated IF signal is coupled from L1 to L2 in two ways: through transformer action and through C2 to the center tap on the coil.

In effect, L3 is connected across the primary winding L1 so that the frequency modulated IF input appears across it at all times. The voltage across L3 is applied to the anode of both diodes, and at the center frequency the diodes conduct equally. When CR1 and CR2 conduct equally, the same quantity of current is passing through both R1 and R2. This current is in opposite directions, and it develops a zero output. The diodes also have a voltage applied as a result of the signal being transformer coupled to L2. This voltage from L2 is 180 degrees out of phase on the two anodes. The diodes then have two voltages applied, and the level of conduction of each diode is controlled by the resultant of the two voltages.

When our signal at the top of L1 swings positive, the voltage on L3 places a positive on both anodes. At the same time, the voltage on L2 places a positive on the anode of CR1 and a negative on the anode of CR2. The resultant voltage causes CR2 to cut off while CR1 conducts. The current through CR1 comes from ground through R1 and C4. This current charges C4 with the polarity indicated.

When the top of L1 swings negative, our situation reverses. CR1 will be cut off and CR2 will conduct only slightly. After a few cycles, C4 and C5 will be equally charged. Since these charges are series opposing and the output is taken across both, our output is zero. This condition will prevail only at the IF signal's center frequency.

Above the center frequency, the reactance of L2 and L3 increases while the reactance of C2 and C3 decreases. The voltage across L3 increases, and it appears on the anodes slightly ahead of the induced voltage from L2. Also, the amplitude of the induced voltage is less. The overall result is a positive output because C4 is charged to a higher level than C5. The level of this difference voltage is varying with the frequency.

Below the center frequency, our capacitive reactance increases and the inductive reactance decreases. This gives us a greater charge on C5, which results in a negative output. The charge and discharge action of C4 and C5 removes all IF components while establishing the amplitude and polarity of our audio output.

The Foster Seeley discriminator is very sensitive to amplitude changes in the input. For this reason, the preceding stage is generally a limiter that removes all amplitude modulation from the IF signal. There is another type of FM demodulator that is relatively insensitive to these amplitude changes; it is commonly called a ratio detector.

Ratio Detector At first glance we may find a close resemblance between the ratio detector and the Foster Seeley discriminator. The ratio detector is also a widely used FM demodulator. The ratio detector is illustrated in Fig. 13–14. Compare this schematic to that of the Foster Seeley discriminator in Fig. 13–13.

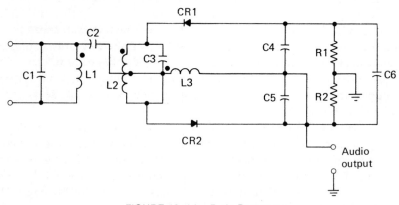

FIGURE 13–14. Ratio Detector

You should have noted three distinct differences between the two schematics: CR1 has been reversed, C6 has been added, and the output is taken from a different point. We now have a path for current from the top of L2 through CR1, through R1, through R2, through CR2, back to the bottom of L2. At the center frequency both diodes conduct equally. They conduct on the negative alternations of the IF signal. The filter section of this demodulator is a bridge circuit, which has a change in voltage distribution with a change in frequency. Figure 13-15 illustrates the voltage distribution on the bridge at the center frequency. These voltages are intended as examples only; they *are not* meant to portray actual values.

Capacitor C6 charges when the diodes conduct and loses very little charge when they are cut off. C4 and C5 also charge as indicated, and their total charge is equal to the charge on C6. The output between ground and the junction of C4 and C5 is zero when these charges are equal.

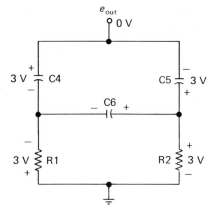

FIGURE 13-15. Center Frequency Voltages

When the signal frequency increases or decreases from center, one of the diodes will conduct more while the other conducts less. This will alter the charge balance of C4 and C5 but will have no effect on C6, nor will it affect the voltage across R1 and R2. The voltage distribution above the center frequency is illustrated in Fig. 13-16.

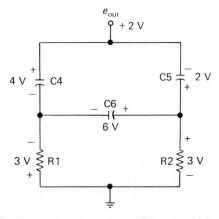

FIGURE 13-16. Above Center Frequency Voltages

Now C4 has charged to a higher level, and C5 has lost some of its charge. The difference between the two charges is +2 V, and that is our output signal amplitude. This condition will reverse when the signal frequency is below center, as illustrated in Fig. 13–17.

Now we have increased the charge on C5 and decreased it on C4. The difference between the two charges is now −2 V. This negative 2 V is the amplitude of our audio output at this time. The charge on C4 and C5 continue to vary with the frequency of the modulating signal.

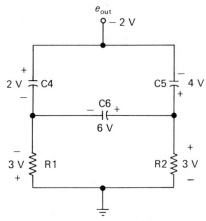

FIGURE 13–17. Below Center Frequency Voltages

AUDIO REPRODUCTION

When our demodulator has extracted the intelligence from the IF signal, we have reproduced the modulating wave that is being used in the transmitter. Now we need a device that will convert the modulating wave back into sound. If all is well, this sound will be the same as the sound producing the modulating wave in the first place. This audio reproduction device is called a speaker. We have speakers in many sizes and shapes ranging from a headphone to large public address speakers.

Magnetic Headphone One of the simplest devices for converting audio signals into sound waves is the magnetic headphone. This headphone is illustrated in Fig. 13–18.

The headset consists of a permanent magnet, two pole pieces, two coils, and a metal diaphragm. The coils are wound around the pole pieces of the magnet, and the audio signal is coupled to the two coils. When we have no input, the permanent magnet exerts a steady

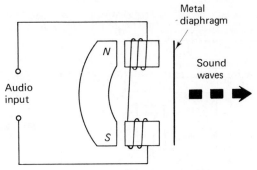

FIGURE 13-18. Magnetic Headphone

force of attraction on the metal diaphragm. The audio signal causes current through the coils, and they become small electromagnets. The fields from the electromagnets either add to or subtract from the strength of the permanent magnet. This sets up a vibration in the metal diaphragm. The vibrating diaphragm produces sound waves with amplitude and frequency similar to the amplitude and frequency of the audio signal. Most headphones are adequate reproducers in the voice range of audio frequencies, but they have poor response to the lower audio frequencies.

Permanent Magnet Speaker The permanent magnet speaker is suitable for many uses including the loudspeaker in a public address system. It consists of a permanent magnet that is mounted on soft iron pole pieces, a voice coil electromagnet, and a speaker cone that is connected to the voice coil. A diagram of this speaker is illustrated in Fig. 13-19.

The audio input to this speaker has been amplified in both voltage and power. The voice coil is wound around a portion of the speaker cone that fits over the center portion of the soft iron pole pieces. This places the coil in the air gap, where the magnetic field from the permanent magnet is very strong. As the audio signal swings positive and negative, it energizes the electromagnet first one way then the other. The interacting magnetic fields cause the electromagnet to move back and forth in the air gap. The distance of movement is controlled by the audio amplitude, the direction of movement is controlled by the audio polarity, and the rate of movement is controlled by the audio frequency. The end result is a vibration, which produces the sound equivalent of our audio signal.

The frequency response of most permanent magnet speakers is excellent at low audio frequencies but tends to drop off at the higher audio frequencies. This can be solved by a more expensive, specially designed speaker or by using two speakers. If we use two

324 Demodulators

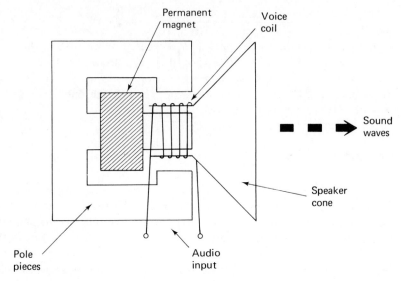

FIGURE 13-19. Permanent Magnet Speaker

speakers, one is standard for low-frequency response, and the other is designed for good response at high audio frequencies.

Dynamic Speaker The dynamic speaker simply uses an electromagnet instead of a permanent magnet. Otherwise, it is very similar to the permanent magnet speaker that we just discussed. This speaker is illustrated in the diagram of Fig. 13-20.

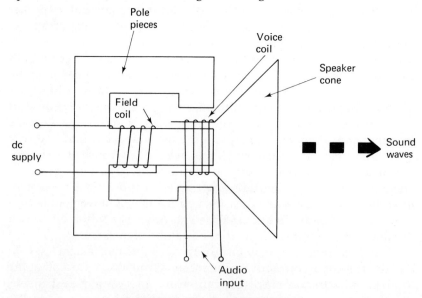

FIGURE 13-20. Dynamic Speaker

Here we have a field coil wound on the center portion of an E-shaped piece of soft iron. Current from a dc power supply energizes the field coil, and it becomes an electromagnet. The soft iron then become pole pieces. The speaker coil is mounted in the same way and reacts in the same way as we discussed for the permanent magnet speaker. The quality and frequency response also closely resemble those of the permanent magnet speaker.

REVIEW EXERCISES

1. Differentiate between demodulation and demodulator.
2. What are the four basic functions of a receiver?
3. What is the significance of a receiver's selectivity rating?
4. Define heterodyning.
5. What is a superheterodyne receiver?
6. A 1200-kHz carrier wave is AM modulated by a single 10-kHz tone. What are the sideband frequencies?
7. The signal of item 6 is mixed with a local oscillator frequency of 1700 kHz. Identify six new frequencies that will be generated.
8. In order to retain the full intelligence, what would be the filter bandpass for the frequencies of item 7?
9. In most commercial broadcast receivers, what is the IF:
    a. Center frequency?
    b. Bandpass?
10. What happens to the intelligence when we pass the AM carrier wave through a mixer stage?
11. What is the difference between a detector and a demodulator?
12. How is AM demodulation accomplished?
13. Explain why rectification is essential to extracting intelligence from an AM signal.
14. What is accomplished by the filtering action in an AM demodulator?
15. How does the audio signal from a demodulator compare to the modulating signal in the transmitter?
16. What is a converter?
17. Describe the two inputs to a mixer.
18. What is the relationship between the carrier wave center frequency and the local oscillator frequency?

Items 19 through 27 refer to Fig. 13-21 (page 326).

19. Identify the stage by name and type.
20. What are the two major objections to this particular type of injection?

326 Demodulators

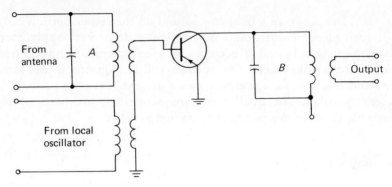

FIGURE 13-21. A Stage in an AM Receiver

21. Draw the waveshapes for both of the inputs and the output.
22. As a rule, we would expect the resonant frequency of tank B to be what frequency?
23. What is done to the circuit when one is tuning in a particular station?
24. What is the relationship of the resonant frequency of Tank A to the center frequency of the received signal?
25. What is the relationship of the local oscillator frequency and the resonant frequency of tank A?
26. Alter this schematic to eliminate both the problems mentioned in item 20. Keep the PNP and common emitter configuration and show both biasing and stabilization circuits.
27. Identify the circuit that you have just drawn both by name and type.
28. Identify the circuit in Fig. 13-22 as to name and type.

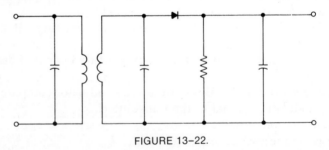

FIGURE 13-22.

29. Draw the input and output waveshapes for Figure 13-22.
30. Draw a schematic for a shunt diode AM demodulator.
31. What are the primary advantage and disadvantage of using a transistor in an AM demodulator?

32. What three functions are performed by a transistor AM demodulator?
33. What is the difference between an FM demodulator and a discriminator?
34. Briefly describe what takes place during FM demodulation.
35. What is the difference between an audio reproducer and a speaker?
36. Describe the function of a speaker.
37. How does the sound from a speaker compare to the sound entering the microphone of the transmitter?
38. Why is a headphone unsuitable for music?
39. How does the frequency response of a permanent magnet speaker correspond to that of a headphone?

Bibliography

Corning, John J., *Transistor Circuit Analysis and Design*. Englewood Cliffs, N.J.: Prentice-Hall, Inc., 1965.

Cutter, Phillip, *Electronic Circuit Analysis*. New York: McGraw-Hill, 1960.

General Electric, *Transistor Manual*. Syracuse, N.Y.: General Electric Co., 1968.

Ghausi, Mohammed S., *Electronic Circuits*. New York: Van Nostrand Reinhold Co., 1971.

Herrick, Clyde N., *Oscilloscope Handbook*. Reston, Va.: Reston Publishing Co., 1974.

Kiner, Milton S., *Transistors in Radio, Television, and Electronics*. New York: McGraw-Hill, 1959.

Lenk, John D., *Handbook of Simplified Solid State Circuit Design*. Englewood Cliffs; N.J.: Prentice-Hall, Inc., 1971.

Lenk, John D., *Handbook of Practical Electronic Tests and Measurements*. Englewood Cliffs, N.J.: Prentice-Hall, Inc., 1969.

Lo, Arthur W., Endres, Richard O., Jawels, Jacob, Waldhauer, Fred D., and Cheng, *Transistor Electronics*. Englewood Cliffs, N.J.: Prentice-Hall, Inc., 1961.

Mandl, Matthew, *Fundamentals of Electronics*, 2nd ed. Englewood Cliffs, N.J.: Prentice-Hall, Inc., 1965.

Marcus, Abraham, *Electronics for Technicians*. Englewood Cliffs, N.J.: Prentice-Hall, Inc., 1969.

Puller, Keats A., Jr., *Conductance Design of Active Circuits*. New York: John F. Rider Publisher, 1959.

RCA Institute, *Transistor Servicing Guide*. Harrison, N.J.: RCA, 1968.

Robinson, Vester, *Handbook of Electronic Instrumentation, Testing, and Troubleshooting*. Reston, Va.: Reston Publishing Co., 1974.

Robinson, Vester, *Electronic Concepts*. Reston, Va.: Reston Publishing Co., 1972.

Robinson, Vester, *Basic Principles of Electronics*. Reston, Va.: Reston Publishing Co., 1973.

Schwartz, Seymour, *Selected Semiconductor Circuits Handbook*. New York: John Wiley & Sons, 1971.

Shockley, W., *Electrons and Holes in Semiconductors*. Princeton, N.J.: D. Van Nostrand Co., 1950.

Shore, B. H., *Fundamentals of Electronics*, 2nd ed. New York: McGraw-Hill, 1970.

Wedlock, Bruce D., and Roberge, James K., *Electronic Components and Measurements*. Englewood Cliffs, N.J.: Prentice-Hall, Inc., 1969.

Answers to Review Exercises

CHAPTER ONE

1. Conductors, semiconductors, and insulators.
2. Conductors have many free electrons, and insulators have very few free electrons.
3. It is a material that is neither a good insulator nor a good conductor.
4. The conductors have the lowest resistivity. Insulators have the highest resistivity. Semiconductors have a resistivity in the middle region between conductors and insulators.
5. (a) Silicon and germanium.
   (b) Single crystal.
6. Eight.
7. Both silicon and germanium have four electrons in the outer shell of each atom. Each atom shares its four valence electrons with adjacent atoms. This gives each atom the effect of having a full valence band of eight electrons.
8. Trivalent materials have three valence electrons in each atom. Pentavalent materials have five valence electrons in each atom.
9. (a) A material with negative carriers has full valence bands with surplus electrons floating about as free electrons.
   (b) A material with positive carriers has a shortage of electrons in the valence band. Each missing electron leaves a hole, which is a small positive charge.
10. The donor materials have a surplus of electrons. The acceptor materials have a surplus of holes (shortage of electrons).
11. Dope the melt with a small quantity of pentavalent material.
12. Dope the melt with a small quantity of trivalent material.
13. P-type material is acceptor material. It has positive current car-

riers because it has a surplus of holes. N-type material is donor material. It has negative current carriers because it has a surplus of electrons.

14. Electron current is the flow of electrons. In an external circuit, electrons leave the negative terminal of the battery and return to the positive terminal. Conventional current exists in the opposite direction to electron flow.
15. The holes in the crystal are repelled by the positive terminal of the battery and attracted by the negative terminal. The electrons composing electron current move in the opposite direction to the movement of the holes.
16. The electron carriers are repelled from the negative terminal of the battery and attracted by the positive terminal of the battery. Electrons composing the electron current move in the same direction as the carriers.
17. By chemically joining a section of P-type material to a section of N-type material in such a manner that they become one continuous crystal.
18. It is an area of stable crystal to either side of the junction.
19. Forward bias aids the movement of both positive and negative carriers and reduces the width of the depletion region. Reverse bias opposes the movement of both positive and negative carriers and increases the width of the depletion region.

20.

21.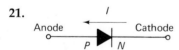

22. It is a junction characteristic that functions as a small internal battery across the junction. The positive side is to the N-type material and the negative side is to the P-type material.
23. (a) Series aiding. (b) Series opposing.
24. (a) It decreases to a negligible level.
    (b) It increases to equal the external bias potential.
25. The carriers are driven at a high velocity and release many additional carriers by bombardment. The action accumulates until an arc occurs through the junction from P- to N-type material. This results in a large reverse current.

332 Answers to Review Exercises: Chapter One

26. It damages the junction and destroys the rectifying qualities of the diode.
27. The quantity of carriers increases as the temperature increases.
28. When junction temperature reaches a certain critical value, more and more carriers are released and temperature continues to rise. If allowed to continue, the diode will become unstable and may be destroyed.
29. The signal diode has a very thin junction, whereas the power supply diode has a much thicker junction. The signal diode must be able to react to small voltages at high speeds and high frequencies. The power supply diode must be able to withstand high currents with minimum resistance.
30. 25°C.
31. The level of direct voltage that will produce the best operating conditions.
32. Forty-five volts of direct voltage in the reverse direction will cause junction breakdown at a temperature of 25°C.
33. The maximum peak value of reverse voltage that can be applied to a power supply diode without danger of breakdown.
34. The maximum forward peak current at a frequency of 60 Hz.
35.

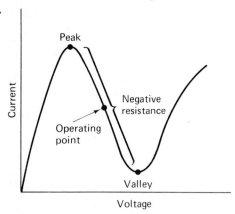

36. The negative resistance portion.
37.

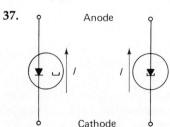

Answers to Review Exercises: Chapter One

38.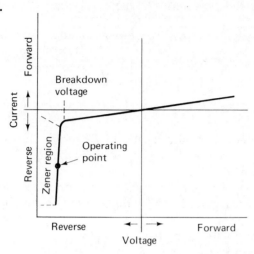

39.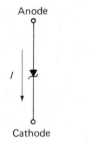

40.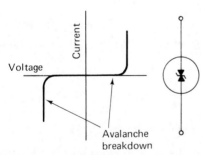

41.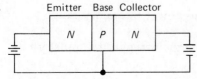

42.

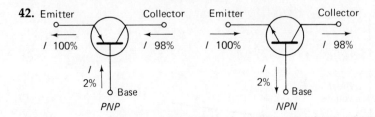

43. The source must be negative with respect to the drain. The drain current is increased or decreased according to the signal variation on the gate.

44. The P channel has P-type material between source and drain with N-type material as a gate. The N channel has the materials in reverse order.

45. Metal oxide semiconductor field effect transistor.

46. The gate potential induces a conduction channel between the source and the drain.

47.

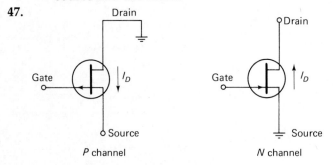

48. A low resistance conduction channel is diffused into the substrate between the source and the drain.

CHAPTER TWO

1. They must be capable of delivering a relatively high current with a low voltage.
2. 16 to 16,000 Hz.
3. One specific frequency in the power frequency range.
4. 95 percent.
5. It was dissipated through energy losses in the transformer.

6. Copper losses — due to wire resistance.
   Hysteresis losses — due to residual magnetism in the core.
   Eddy current losses — due to circulating currents induced into the core.
7. 12:1.
8. 8 mA.
9. The maximum peak voltage that can be applied across the diode in the reverse direction.
10. (a) It increases the PIV rating.
    (b) It increases the current capacity.
11. (a) High voltage and low current.
    (b) Low voltage and high current.
12. To change alternating voltage to pulsating direct voltage.
13. (a) Same as the input frequency.
    (b) Twice the input frequency.
14.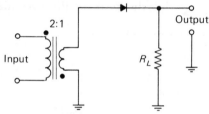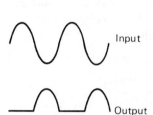
15. 60 V peak to peak on the secondary and 30 V maximum output from the rectifier.
16. The maximum output from the rectifier is one-half the amplitude of the secondary peak. This happens because each of the two diodes uses only half the turns on the transformer's secondary winding.
17. Four: one for each phase and one common for ground. If all three phases are perfectly balanced, the ground wire may be eliminated.
18. It is to reduce the amplitude of the ripple in the pulsating direct voltage.
19. This filter is adequate when a half-wave rectifier is powering a circuit that draws very little current.
20. It blocks ac and slows the capacitor discharge. The coil is essential when the load current becomes reasonably large.
21. An L-type, choke input, LC filter.
22. This type of load circuit requires a very high voltage but draws practically no current.

23.

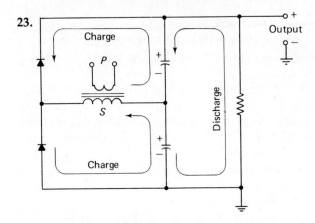

24. 570 V. The peak voltage on the primary is 1.14 × 50, or 57 V. The secondary peak is 5 × 57 V, or 285 V. The voltage across the load resistor is equal to twice the peak of the secondary voltage, and 2 × 285 V is 570 V.

25.

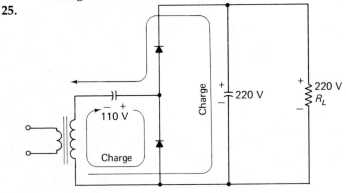

26. Voltage divider.

27. (a) A rheostat is a current control. It is composed of coils of low-resistance, large-diameter wire. It is a variable resistor that uses two contacts. It is connected in series with the load circuit.

(b) A potentiometer is a voltage control. It is a variable resistor with a high resistance value. It uses three contacts. The total value parallels the power supply, while the center arm taps off voltage for the load circuit.

28. Changes in input voltage levels and changes in output load current.

29. The shunt regulator has a variable component in parallel with

the load circuit. The series regulator has a variable component in series with the load circuit.
30. An electronic series regulator.
31. It provides a constant regulated output voltage despite all reasonable changes of input voltage, output current, and temperature. The negative temperature coefficient of two junction diodes balances the positive temperature coefficient of the Zener diode. The result is a constant value of shunt resistance over a wide change in temperature.
32. Greater stability, closer tolerances, and a means of adjusting the output level.
33.

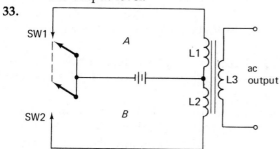

34.

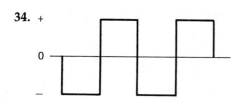

35. A dc to dc converter is simply a dc to ac converter with a rectifier circuit connected to the transformer secondary.

CHAPTER THREE

1. It increases the magnitude of a signal.
2. Voltage, current, and power.
3. Reverse the polarity of all bias voltages.
4. (a) Class A. (b) Class B. (c) Class AB.
5. The type of signal being amplified and the desired shape of the output.
6. (a) Common emitter. (b) Common base. (c) Common collector.

7.

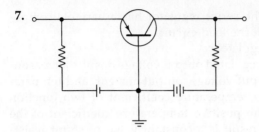

8. About 10 watts.
9. Alpha is associated with the common base configuration. It is the current amplification factor.
10. 0.87.
11. (a) Voltage and power. (b) Voltage, current, and power. (c) Current and power.
12. (a) In phase. (b) 180 degrees out of phase. (c) In phase.
13. The frequency that reduces the current gain to 0.707 of its low-frequency value.
14. The diffused capacitances across the junctions become more of a short as signal frequency goes higher.
15. (a) The common emitter and the common collector have about the same current gain. (b) The common base. (c) The common collector.
16. The values of base current and collector current for all possible values of collector voltage.
17. 40.
18. (a) $\beta = \dfrac{\alpha}{1 - \alpha}$ (b) $\alpha = \dfrac{\beta}{1 + \beta}$
19. 19.
20. 1500.
21. High input impedance and low output impedance.
22. Matching impedance between a high and a low impedance circuit.
23. $I_e = I_c + I_b$.
24. 41.
25. Because power gain is $A_i \times A_v$ and A_v is always slightly less than one.
26. Cutoff and saturation.
27. The transistor is cut off when all current from emitter to base ceases.
28. Both input and output are forward-biased.
29. (a) Set the bias so that no part of the input signal will ever reduce the forward bias to zero. (b) Set the bias so that the signal can never cause both junctions to be forward-biased.

Answers to Review Exercises: Chapter Four

30. Halfway between cutoff and saturation.
31. It causes an increase in leakage current.
32. Emitter, degenerative feedback.
33. (a) Improves stability. (b) Decreases gain.
34.

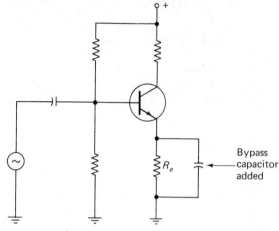

35. Collector, degenerative feedback.
36.

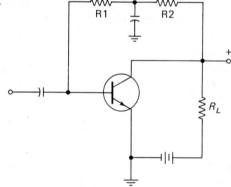

CHAPTER FOUR

1. (a) Maximum capabilities. (b) Performance in a given circuit.
2. Load lines and equivalent circuits.
3. Equivalent circuits. The small signals cause so little change that the load line method is not reliable.
4. A quantity whose value varies with the circumstances of its application.

5. Input current, output current, input voltage, and output voltage.
6. Because the transistor amplifies both voltage and current.
7. (a) Single circle. (b) Double circle.
8. The quantities are measured under specified open-circuit conditions.
9. r_{ib}, r_{rb}, r_{fb}, and r_{ob}.
10. r_i = input resistance with output open.
    r_r = reverse transfer resistance with input open.
    r_f = forward transfer resistance with output open.
    r_o = output resistance with input open.
11. The third letter (or second subletter).
    (a) b. (b) e. (c) c.
12. (a) r_{ie}. (b) r_{re}. (c) r_{fe}. (d) r_{oe}.
13. Conductance.
14. g_{ie}, g_{re}, g_{fe}, and g_{oe}.
15. g_i = input conductance with output shorted.
    g_r = reverse transfer conductance with input shorted.
    g_f = forward transfer conductance with output shorted.
    g_o = output conductance with input shorted.
16. (a) g_{ie}. (b) g_{re}. (c) g_{fe}. (d) g_{oe}.
17. Hybrid means mixed. A hybrid parameter is developed by using one voltage and one current as independent variables, so it is a mixture of open- and short-circuit parameters.
18. h_{ie}, h_{re}, h_{fe}, and h_{oe}.
19. h_i = hybrid input resistance.
    h_r = hybrid reverse voltage gain.
    h_f = hybrid forward current gain.
    h_o = hybrid output conductance.
20. (a) h_{ie}. (b) h_{re}. (c) h_{fe}. (d) h_{oe}.
21. h_{11} = input resistance.
    h_{12} = reverse voltage gain.
    h_{21} = forward current gain.
    h_{22} = output conductance.
    α_f = forward current gain.
    μ_r = reverse voltage gain.

CHAPTER FIVE

1. An amplifier with a large signal amplitude as compared to the bias on the circuit.
2. It is largely a matter of choice. When a signal produces a change

Answers to Review Exercises: Chapter Five 341

large enough for accurate graphic analysis, the amplifier may be classified as a large signal amplifier.
3. Load resistance and output bias.
4. All dynamic characteristics of a particular transistor in a given set of operating conditions.
5. The points of maximum V_c with zero I_c and maximum I_c with zero V_c.
6. It is the quiescent point of input current, output voltage, and output current. It is established by setting the input bias for the desired level of dc input current.
7.

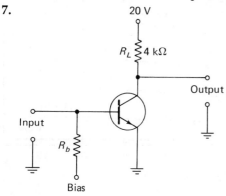

8. 19,600 ohms.
9.

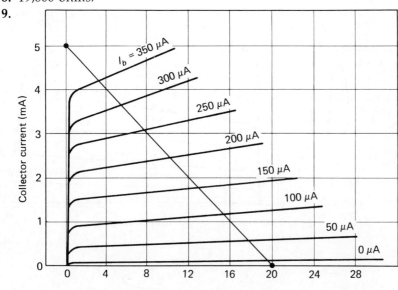

10. $I_b = 200$ μA, $I_c = 2.4$ mA, $V_c = 10.4$ V. Operating point as shown in Fig. 14–21.

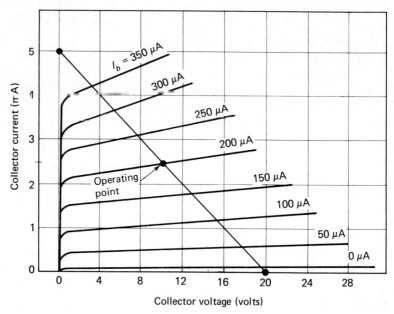

11.

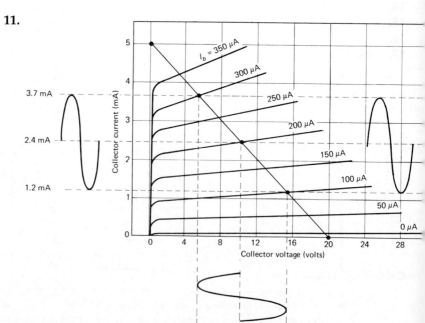

Answers to Review Exercises: Chapter Six 343

12. (a) $V_{in} = I_b \times R_{in}$
$= 200 \ \mu A \times 400 \ \Omega$
$= 80 \ mV$

(b) $A_r = \dfrac{V_{out}}{V_{in}}$
$= \dfrac{10 \ V}{80 \ mV}$
$= 125$

(c) $A_i = \dfrac{I_{out}}{I_{in}}$
$= \dfrac{2.5 \ mA}{200 \ \mu A}$
$= 12.5$

(d) $G = A_r \times A_i$
$= 125 \times 12.5$
$= 1562.5$

13. Bipolar.
14. A common base amplifier.
15. 40 mW.
16. 31.25 mW.
17.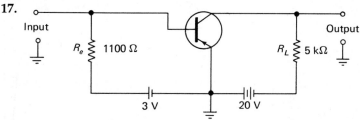
18. 0.25 V.
19. 12.5 V.
20. (a) 0.3 V. (b) 0.9 mA to 3.8 mA. (c) 6 V to 15.5 V.
21. (a) 0.967. (b) 31.67. (c) 30.6.

CHAPTER SIX

1. All equipments that produce audible sound. Public address systems, radio receivers, and sound recorders and reproducers.

2. Any device that provides the input signal. It may be a signal generator, an antenna, or a previous stage.
3. Preamplifiers, drivers, and power amplifiers.
4. All the quiescent conditions of current and voltage.
5. The ac signal rides the quiescent values.
6. Preamplifier.
7. Signal level divided by noise level.
8. The input signal-to-noise ratio divided by the output signal-to-noise ratio.
9. Source resistance and signal frequency.
10. Place a resistor in series with the base or in series with the emitter.
11. RC, transformer, impedance, and direct.
12. A gain adjustment in an audio amplifier.
13. The tone control varies frequency response as the volume control varies gain.
14.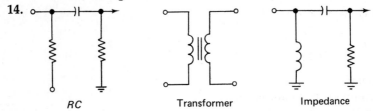

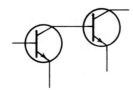

15.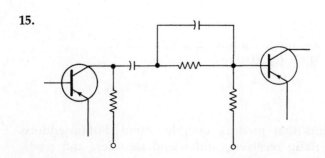

16.

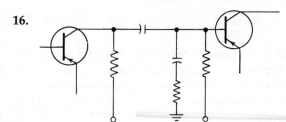

17. (a) 795 kΩ. (b) 795 Ω. (c) 39.75 Ω.
18. Transformer.
19. The signal is developed across the coil, which is a virtual short for very low frequencies.
20. A transformer responds only to ac, and all other types of coupling use a series capacitor that blocks the dc.
21. (a) Volume control. (b) Low-frequency tone control. (c) High-frequency tone control.
22.

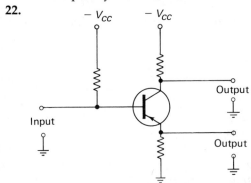

23. It provides more driving power.
24. They deliver high output power with high power efficiency.
25. Slow changeover on a nonlinear portion of the curve. It is prevented by a small forward bias on both transistors.
26.

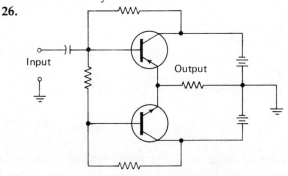

CHAPTER SEVEN

1. It is an amplifier that is tuned to amplify a narrow band of frequencies and suppress all other frequencies.
2. It is a descriptive term of how well an amplifier does its job of amplifying the frequencies in a chosen band and eliminating all other frequencies.
3. Primarily the interstage coupling.
4. Parallel resonant circuits.
5. The series resonant circuit rejects a band of frequencies and is inflexible for impedance transfer.
6. It matches the output impedance of one stage and the input impedance of the other.
7. $X_L = X_C$, line current in phase with the applied voltage, and maximum current circulating in the tank.
8. A low Q tank circuit.
9. The resonant frequency is the center of the band that is passed.
10. 460 kHz.
11. 66.7 kΩ.
12. 115.6.
13. 4 kHz.
14. 458 kHz to 462 kHz.
15. (a) No effect on frequency.
    (b) Impedance is reduced by 50 percent.
    (c) Q is halved.
    (d) Bandwidth is doubled.
    (e) Bandpass is now 456 kHz to 464 kHz.
    (f) Selectivity is reduced.
16. 34.6 mA.
17. 1 mA.
18. 34.6.
19. 13.3 kHz.
20. (a) No effect on tank current.
    (b) Line current halves.
    (c) Q doubles.
    (d) Bandwidth halves.
21. By selecting the proper primary to secondary turns ratio.
22. $\dfrac{N1}{N2} = \sqrt{\dfrac{R_o}{R_i}}$
23. A transformer with a single winding with taps to determine which portion of the winding will serve as primary and which portion will serve as secondary.

Answers to Review Exercises: Chapter Eight   347

24. A flatter frequency response between half-power points, sharper cutoff outside the half-power points, and more attenuation for frequencies outside the bandpass.
25. Positive or regenerative.
26. The higher the frequency, the greater the feedback.
27. By a degenerative external feedback that is equal to the internal regenerative feedback.
28. The internal feedback has a tendency to cause the amplifier to oscillate.
29. The external feedback cancels both the resistive and reactive components of the internal feedback.
30. When the external feedback circuit cancels only the reactive component of the internal feedback.
31. It keeps amplitude constant regardless of signal strength.
32. A voltage level proportional to signal strength varies the bias on an amplifier to provide high gain for weak signals and low gain for strong signals.
33. Control of emitter current and control of collector voltage.

CHAPTER EIGHT

1. The video signal is composed of a fundamental frequency and many harmonics. The narrow band amplifier does not respond to enough frequencies to reproduce the video signal.
2. (a) A single fundamental frequency.
   (b) A fundamental frequency and an infinite number of odd harmonics.
3. (a) A few kHz. (b) A few kHz. (c) A few kHz. (d) A few MHz.
4. Pulses, square waves, sawtooth waves, and stepped waves.
5. Phase distortion and amplitude distortion.
6. A three-winding transformer that is especially designed for video signals.
7. Their frequency response curve is too narrow, and they will not function at the high end of the video band.
8. (a) Low frequencies are attenuated.
   (b) High frequencies are shorted out.
9. The coupling capacitor attenuates the low frequencies, while the distributive capacitance shunts the high frequencies to ground.
10. The signal is divided between the X_C of the coupling capacitor and the resistance of the junction. Only the portion across the junction resistance serves as input.

11. The higher the value of the load resistor, the greater the amplification and the narrower the frequency response. A low-value load resistor broadens the response but reduces the amplification.
12. There would be no significant effect. R_b is in parallel with R_j. R_j has a maximum resistance of 1500 ohms. When $R_b = 10R_j$, a further increase in R_b has no appreciable effect.
13. 1154 ohms.
14. (a) 1363 ohms. (b) 1456 ohms.
15. (a) 5300 ohms. (b) 7550 ohms.
16. About 10 percent.
17.

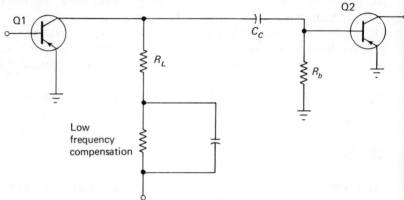

18. The capacitor offers a high impedance to low frequencies, which places the added resistance in series with R_L. The larger load resistance provides greater amplification for the low frequencies.
19.

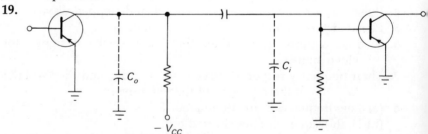

20. The capacitors are parallel to the signal path and become closer to a short as the frequencies increase.

21.

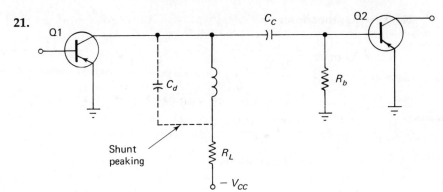

22. R_L should equal X_L and $X_L = X_C$; therefore;

$$R_L, X_L, \text{ or } X_C = \frac{1}{2\pi fC}$$

$$= \frac{0.159}{3.5 \times 10^6 \times 25 \times 10^{-12}}$$

$$= \frac{159 \times 10^3}{87.5}$$

$$= 1817 \text{ ohms}$$

$$L = \frac{X_L}{2\pi f}$$

$$= \frac{1817}{6.28 \times 3.5 \times 10^6}$$

$$= \frac{1817 \times 10^{-6}}{21.98}$$

$$= 82.7 \times 10^{-6}$$

$$= 82.7 \ \mu\text{H}$$

23.

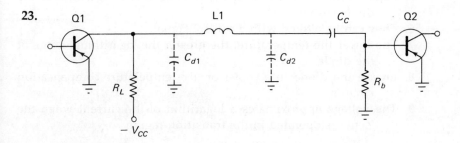

24. $X_L = X_C$; therefore,

$$X_L \text{ or } X_C = \frac{1}{2\pi fC}$$

$$= \frac{0.159}{3.5 \times 10^6 \times 15 \times 10^{-12}}$$

$$= \frac{159 \times 10^3}{52.5}$$

$$= 3028.6 \text{ ohms}$$

Then

$$L = \frac{X_L}{2\pi f}$$

$$= \frac{3028.6}{6.28 \times 3.5 \times 10^6}$$

$$= \frac{3028.6 \times 10^{-6}}{21.98}$$

$$= 138 \ \mu\text{H}$$

25. Place a resistor in parallel with the coil.

CHAPTER NINE

1. Nearly all.
2. Logarithmic and square law.
3. Amplitude compression, amplitude expansion, and frequency response.
4. Either in the electrical resistance or in the transfer characteristics.
5. Logarithmic; the change in diode voltage is a logarithm of the diode current.
6. They can be altered with a control signal.
7. The lower the temperature, the greater the logarithmic range of the diode.
8. Encase the diode in a cooler or use temperature compensation circuits.
9. The voltage approximates a logarithm of the current when the Zener is operated in the transition region.

10.

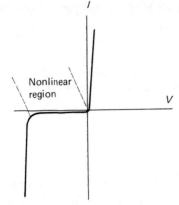

11. Hold the collector voltage at a constant value.
12. Use an external feedback circuit.
13. When the input signal controls either the direct voltage or current.
14. The parameter h_i varies directly with dc, and gain can be made to vary inversely with h_i.
15. The parameter h_f varies directly with dc.
16. The interbase resistance decreases.
17. The silicon in the interbase region is conductivity modulated.
18. (a) 8000 ohms. (b) 100 ohms.
19. At currents between 1 and 5 mA.
20. When operated at voltages below pinchoff.
21. When used in circuits for pulses of short duration.
22. A voltage-sensitive resistor.
23. The input voltage can control the resistance of the varistor, and the varistor current can be used to control the gain of an amplifier.
24. A temperature-controlled resistor.
25. The temperature (and resistance) of the thermistor can be controlled by the input signal, and the thermistor current can be used to control the gain of an amplifier.
26. It is a square law relationship; the output voltage is proportional to the heating effect of the input.
27. (a) From dc to more than 300 MHz.
(b) From 3 mA to 1 A.
28. It is used to generate a logarithmic relationship between the input and output signal.
29. As input voltage amplitude goes up, loading resistance increases, and gain decreases.

352 Answers to Review Exercises: Chapter Ten

30. As input voltage amplitude goes up, loading resistance decreases, and gain increases.
31. The potentiometers.
32. The detector.
33. As the signal amplitude increases, the AGC voltage level in- increases and reduces the gain of the controlled amplifier.

CHAPTER TEN

1. (a) Both supply alternating currents. The mechanical generators are rotating devices that supply currents at frequencies up to a few thousand hertz. The electronic oscillator is a circuit that converts direct current to alternating current. Oscillators can supply ac at frequencies up to and including visible light.
   (b) The mechanical generators are needed to supply ac at low frequencies. The electronic oscillators are needed to supply ac at high frequencies.
2. A dc power supply, an amplifier, a frequency determining device, and regenerative feedback.
3. The oscillator must have a steady source of dc. It is usually taken from a regulator circuit.
4. The common base has a reasonable gain of voltage, a reasonable gain of power, and no phase reversal of the signal.
5. There is a great difference in the input and output impedances, and the current gain is less than unity.
6. The feedback circuit must match the output impedance to the input impedance in order to avoid excessive energy losses.
7. The common emitter has a phase shift of 180 degrees between input and output signals.
8. Input and output impedances are both moderate, and the circuit produces a reasonable gain of current, voltage, and power.
9. The common collector has a wide difference between input and output impedances, and the voltage gain is less than unity.
10. The flywheel effect is the interchange of energy between the electrostatic field of a capacitor and the electromagnetic field of an inductor.
11. When the switch opens, the capacitor forces electrons from its negative side through the coil to its positive side until the capacitor is discharged. The field on the coil then takes over and keeps current in the same direction until the capacitor is charged in the opposite direction. This terminates one alternation. The next alternation is identical, except that the direction of current is reversed.

12.

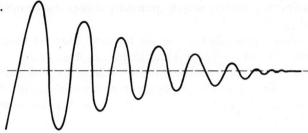

13.

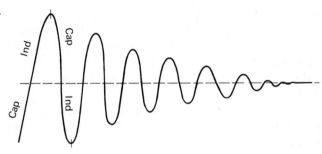

14. Energy must be periodically injected in order to replace the losses of the circuit.
15. The values of L and C.
16. 3180 Hz.
17. For stable frequency the operating point must be on a linear portion of the transistor characteristic. Nonlinear operation changes the transistor's parameters and shifts the frequency.
18. A change in temperature has a tendency to move the operating point with a resultant shift in frequency.
19. We can connect an external capacitor across the collector-emitter electrodes.
20. They divide the $-V_{CC}$ to provide proper input bias for the amplifier.
21. It is a swamping resistor to provide temperature stability.
22. C2 is an ac bypass to shunt the signal around R3. This prevents signal degeneration.
23. L2 is the inductor portion of the frequency determining tank circuit and the primary winding for the transformer. L1 is a tickler coil pickup loop for feedback. L3 is the output winding of the transformer.
24. C3 and L2.
25. By varying the value of C3.
26. (a) Armstrong.
    (b) The tickler coil feedback arrangement.
27. Feedback ceases and decreases the forward bias. This reverses

354 Answers to Review Exercises: Chapter Eleven

the feedback polarity, which gradually drives the transistor toward cutoff.
28. Feedback ceases and allows the dc forward bias to become effective. This starts current and begins another cycle.
29. The feedback is tapped from the capacitor side of the tank circuit.
30. The means of tapping feedback energy from the tank circuit. The Colpitts uses tapped capacitors while the Hartley uses a tapped inductor.
31. Greater power output. The push-pull amplifier drives the tank circuit all the time and produces greater power.
32. It is the quality of certain crystals that enables them to produce a voltage under mechanical stress or mechanically to vibrate when voltage is placed across them.
33. They produce a frequency with a high degree of stability.
34. Series mode and parallel mode.
35. The impedance of the circuit into which the crystal is connected.
36. The tank circuit can be tuned to resonate at one of the harmonics of the fundamental frequency of the crystal.
37. This is usually accomplished by having several different crystals with a common crystal holder. When the crystal is changed, the frequency is changed.
38. One RC combination can shift the phase of a signal up to 90 degrees. Several combinations can be adjusted to shift the phase exactly 180 degrees for the desired frequency.
39. Higher power output and greater stability.

CHAPTER ELEVEN

1. The difference is in the output waveshape. The sinusoidal oscillator produces a sine wave. The nonsinusoidal oscillator produces an ac output of any shape other than a sine wave.
2. Relaxation oscillators and synchronized oscillators.
3. The relaxation oscillator is free-run. It requires no input, and it supplies a continuous string of output signals. The synchronized oscillator provides an output only when it is triggered by a signal from an external source.
4. In most cases, an RC circuit.
5. Five time constants.
6. 1.25 seconds.
7. (a) One time constant.
   (b) Three time constants.
   (c) Five time constants.

8. (a) 0.25 seconds. (b) 0.75 seconds. (c) 1.25 seconds.
9. (a) 8.6 V. (b) Five.
10. The simple relaxation oscillator is the same circuit with an electronic device for a switch. The opening and closing of the switch determines when the capacitor charges and when it discharges. The output is a sawtooth waveshape.
11.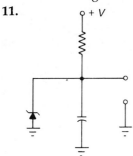
12. Sawtooth.
13. When the Zener diode is cut off the capacitor is charging. When the diode conducts (avalanche breakdown) the capacitor discharges.
14. About the first 10 percent.
15. A slight mismatch in the circuit allows one transistor to saturate and cuts the other off.
16. The value of the RC coupling components.
17. It produces a series of square or rectangular waves, and they can be taken from the collector of either transistor.
18. A series of trigger pulses of short duration.
19. (a) Zero bias when power is applied.
    (b) Positive feedback.
    (c) At saturation feedback ceases and the capacitor discharge cuts it off.
20. Add cutoff bias to the transistor and provide a trigger pulse from an external source.
21. 1:1; we get one pulse out for each pulse in.
22. The frequency of the trigger pulses would have to be slightly higher than the free-run frequency of the oscillator.
23. (a) The frequency of the input trigger. Each input pulse produces one output.
    (b) The RC time constant in the coupling circuit.
24. This is a bistable multivibrator with steering diodes.
25. It has a faster switching action.

356 Answers to Review Exercises: Chapter Twelve

26. They steer the signal so that it is applied only to the base of the cutoff transistor.
27. They provide fast coupling for the feedback to speed the switching action.
28.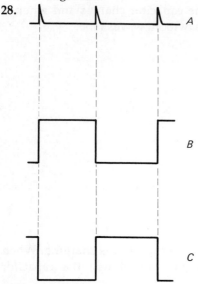
29. Two input pulses produce one cycle of output.
30. The input can be any waveshape that can activate the circuit. The output is a square or rectangular wave.

CHAPTER TWELVE

1. Modulation is the process of impressing an intelligence upon a carrier wave; a modulator is a device that we use to make this impression.
2. Amplitude and frequency.
3. A microphone converts sound waves into electrical energy. The sound wave is intelligence, and the electrical equivalent of the sound wave can be a modulating wave.
4. A direct current is passed through a carbon pile. The diaphragm is arranged so that sound waves striking the diaphragm cause a varying pressure on the carbon granules. The varying pressure causes a varying resistance and a current which is the electrical equivalent of the sound wave.
5. Random changes in the resistance of the carbon granules.

Answers to Review Exercises: Chapter Twelve   357

6. It has a poor frequency response.
7. The microphone has a low output impedance. The amplifier has a high input impedance. The transformer offers the easiest method of balancing or matching these extreme conditions.
8. The piezoelectric effect of Rochelle salt or quartz crystals.
9. The crystal microphone has a high output impedance and can be connected directly into an amplifier.
10. It has a low output, which may require several stages of amplification.
11. A moving coil in a permanent magnetic field.
12. An electromagnetic wave that carries the intelligence from the transmitter to the receiver.
13. An intelligence signal that is impressed upon a carrier wave.
14. In amplitude modulation the intelligence is impressed into the amplitude of the carrier wave. In frequency modulation the intelligence is impressed into the frequency of the carrier wave.
15. (a) The upper sideband is the sum of the frequencies of the carrier wave and the modulating wave.
    (b) The lower sideband is the difference between the frequencies of the carrier wave and the modulating wave.
16. Subtract the lower sideband frequency from the upper sideband frequency.
17. (a) 210 kHz. (b) 190 kHz.
18. 20 kHz.
19. It is the ratio of modulating wave voltage to carrier wave voltage.
20. (a) 0.9. (b) 90 percent.
21. (a) 0.905. (b) 90.5 percent.
22. (a) Waveshape b is overmodulation.
    (b) Waveshape c is undermodulation.
    (c) Waveshape a is correct modulation.
23. The amplifier is used to amplify the carrier wave, and the modulating wave is used to vary the gain of the amplifier. This action causes the amplitude of the carrier wave to vary at the rate of the modulating wave.
24. Base injection, emitter injection, and collector injection.
25. Base and emitter injection are low-level. Collector injection can be either low- or high-level.
26. Low-level modulation deals with low amplitude waves. The output carrier wave may require several stages of amplification before transmission. High-level modulation takes place at a later point in the RF amplifier chain. The signals are high in amplitude, and the carrier wave may be broadcast without

further amplification. Low-level modulators operate class A, and high-level modulators operate class C.
27. The oscillator is generating a carrier wave, but the gain of the transistor is varied by the modulating wave. This impresses the modulating wave into the amplitude of the carrier wave while it is being generated.
28. (a) Usually nothing. The FM is slight and can be tolerated.
    (b) We use a limiter to remove the AM from the carrier wave.
29. Deviation is the amount of change of a carrier wave frequency to either side of the center frequency.
30. Rate of deviation is the number of complete deviation cycles that occur per second. Degree of deviation is the number of hertz that the carrier wave frequency changes to either side of center.
31. (a) The frequency of the modulating wave.
    (b) The amplitude of the modulating wave.
32. (a) 200.01 MHz and 199.99 MHz.
    (b) 200.02 MHz and 199.98 MHz.
33. When their amplitude is reduced to one percent of the carrier wave amplitude.
34. It is a legal limit set by the Federal Communications Commission.
35. The ratio of deviation to the highest modulating frequency.
36. 5.
37. Locate the modulation index in the modulation index table. The table indicates the number of significant sidebands and the bandwith for each modulation index.
38. The oscillator portion is generating a carrier wave. The modulating wave alters the collector to emitter capacitance of a second transistor. This capacitance change varies the resonant frequency of the frequency determining tank circuit. The frequency of the carrier wave being generated is caused to vary at the same rate as the modulating signal.

CHAPTER THIRTEEN

1. Demodulation is the process of extracting the intelligence from a carrier wave. A demodulator is a device that performs this extraction process.
2. Reception, selection, demodulation, and reproduction.
3. It indicates the receiver's ability to differentiate between desired and undesired signals.
4. Heterodyning is the process of mixing two frequencies together in order to produce a third frequency.

Answers to Review Exercises: Chapter Thirteen 359

5. A receiver that employs the principle of heterodyning.
6. 1210 kHz and 1190 kHz.
7. 2910 kHz, 490 kHz, 2900 kHz, 500 kHz, 2890 kHz, and 510 kHz.
8. 490 kHz to 510 kHz.
9. (a) 455 kHz. (b) 450 to 460 kHz.
10. The intelligence is transferred to the amplitude of the IF signal.
11. No difference. These are two names for the same circuit.
12. By a process of rectification and filtering.
13. The signal has an average value of zero. The signal must be rectified in order for the amplitude variations to be detected.
14. The IF components are removed, and only the audio variations are retained.
15. They should be exact duplicates.
16. A stage that combines the functions of local oscillator and mixer into a single stage.
17. One is a continuous RF from the local oscillator. The other is a modulated carrier wave from the antenna.
18. Generally, the local oscillator frequency is 455 kHz above the center frequency of the carrier wave.
19. It is a mixer, and the type is base injection.
20. The carrier wave frequency interferes with the local oscillator operation, and the local oscillator frequency couples back to and radiates from the antenna.
21.

Modulated carrier wave from antenna

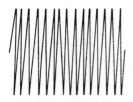
Continuous RF from local oscillator

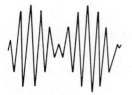
Modulated IF output

360 Answers to Review Exercises: Chapter Thirteen

22. 455 kHz.
23. We vary the resonant frequency of tank A, and we vary the local oscillator frequency.
24. They are the same.
25. The local oscillator is tuned to the center frequency of tank A plus the center of the IF band. This generally places the local oscillator frequency at 455 kHz above the resonant frequency of tank A.
26.

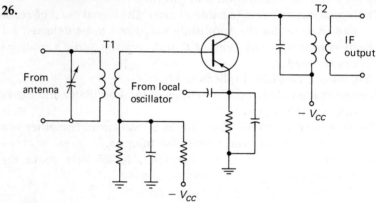

27. It is a mixer, and it is of the emitter injection type.
28. It is an AM demodulator or detector of the series diode type.
29.

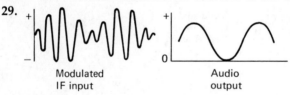

30.

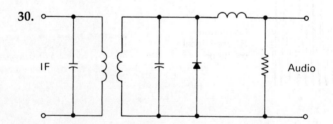

31. The main advantage is the ability to process small signals. The main disadvantage is its inability to handle large signals.

32. Amplification, rectification, and filtration.
33. None; they are two names for the same circuit.
34. The discriminator converts the frequency modulation to a varying amplitude voltage, rectifies, and filters out the IF components. The resulting signal is an audio with frequency according to the rate of modulation and amplitude according to the degree of modulation.
35. None; they are two names for the same device.
36. An audio signal is coupled into the voice coil, and this sets up a vibration in the speaker. The vibrations generate the sound wave equivalent of the audio input signal.
37. If we assume perfect conditions and excellent equipment, the sounds should be exactly the same.
38. They have poor frequency response for the lower audio frequencies.
39. Exactly opposite. Most permanent magnet speakers have a poor response to high audio frequencies.

Index

Ac beta, 76
Acceptor material, 3
Alpha, 71, 111
Ampere, 3
Amplification
  features, 65
  principles, 65
Amplifiers
  audio, 143
  basic, 65
  circuit configurations, 6
    common base, 68, 133
    common collector, 80
    common emitter, 74
    common source, 138
  classes, 67
  current, 65, 67
  large signal, 115
  logarithmic, 229
    automatic gain control, 230
    constant output, 232
    dc amplification, 232
  modulators, 288
  nomenclature, 66
  power, 65, 67
  small signal, 95
  tuned, 174
    coupling, 154, 174, 181, 183
    feedback, 185
  voltage, 65, 67
  wide band, 196
    coupling, 198, 200
    response, 201

Amplitude demodulation, 314
  series diode, 315
  shunt diode, 316
  transistor, 317
Amplitude modulation, 385
  degree, 287
  high level, 291, 292
  ideal, 288
  low level, 290, 292
  percentage, 287, 299
  principles, 285
  sidebands, 286, 298
Armstrong oscillators, 243, 244, 252
Astable multivibrator, 265
Audio
  circuit controls, 159
  drivers, 143, 161
  frequencies, 143
  power amplifiers, 143, 164
  preamplifiers, 143, 149
    noise factor, 149
  push-pull amplifiers, 164
    low bias, 165
    no bias, 165
  reproduction, 322
    dynamic speaker, 324
    magnetic headphone, 322
    permanent magnet speaker, 323
  single ended amplifier, 164
  single-stage phase inverter, 162
  two-stage phase inverter, 163
Automatic gain control, 189

Autotransformer coupling, 182
Avalanche,
  breakdown, 8, 16
  current, 9, 15, 39

Bandwidth
  amplitude modulation, 177
  frequency modulation, 299
Barrier potential, 7, 8
Base
  injection mixer, 309
  injection modulator, 188
  spreading resistance, 186
Basic
  amplifier, 65
  oscillator, 241
Bias
  forward, 6
  input, 20
  output, 20
  reverse, 8
  stabilization, 84
Beta, 75, 111
Bifilar transformer, 198
Bipolar transistor, 1, 18, 19, 65, 124
Bistable multivibrator, 272
Black box, 97
Blocking oscillator, 267, 269
Breakdown
  potential, 8, 15
  voltage, 10
Bridge rectifier, 42

Capacitance coupling, 188
Capacitive rating, 10
Capacitor
  input filter, 44
  microphone, 300
Carbon
  microphone, 282
  pile, 282
Carrier wave, 281, 285, 295
Cascade voltage doubler, 48
Center frequency, 296
Channel
  depletion, 26

  enhancement, 25
  N-, 23, 25
  P-, 23, 25
Characteristic Zener breakdown, 54
Chemical joining, 4
Choke input filter, 46
Classes of amplifiers, 67
Collector injection modulator, 291
Colpitts oscillator
  common base, 245
  common emitter, 246
Common
  base amplifier, 68, 133
    alpha, 71, 111
    bias, 68
    characteristics, 70, 71, 74
    current gain, 137
    cutoff frequency, 73
    input resistance, 72
    load line, 136
    output resistance, 72
    power curve, 136
    power gain, 73, 138
    signal phase relation, 73
    voltage gain, 72, 138
  collector amplifier, 80
    bias, 80
    characteristics, 84
    current gain, 80
    impedance matching, 83
    input resistance, 83
    output resistance, 83
    power gain, 83
    signal phase relation, 82
    voltage gain, 82
  diodes, 1, 4, 5
    classification, 9
  emitter amplifier, 74
    ac beta, 76
    beta, 75, 111
    characteristics, 75, 79
    current gain, 77
    power gain, 79
    relation of alpha and beta, 76
    signal phase relation, 78
    voltage gain, 77
  source amplifier, 138
    avalanche breakdown, 140

load line, 139
N-channel JFET, 138
operating point, 139
power curve, 139
Complementary symmetry, 169
Compression wave, 281, 301
Conductor, 1
Constant
output amplifier, 232
power curve, 129, 136, 139
Constructing
dynamic transfer curve, 124
load line, 116
Conditions of resonance, 175
Control
automatic gain, 189, 230
collector voltage, 189
emitter current, 191
current, 51
tone, 160
treble, 161
voltage, 50
volume, 159
Converters, 59, 309, 313
dc to ac, 59
dc to dc, 61
Converting parameters, 107
Copper losses, 32
Coupling
audio amplifier
direct, 158
impedance, 157
RC, 154
transformer, 156, 181
tuned amplifier, 174, 181, 182, 183
Covalent bonding, 2
Crystal
controlled oscillators, 250
Armstrong, 252
Colpitts, 253
microphone, 183
Crystalline structure, 2
Cube root function, 222
Current
amplifier, 65, 67
carriers, 2, 5, 6
conventional, 3
dynamic, 147

gain parameters, 112
electron, 3
in a transistor, 21
quiescent state, 144
—voltage distribution, 143
Zener, 54

Dc to ac converter, 59
Dc amplifier, 232
Dc to dc converter, 61
Decreasing attenuation, 205
Degenerative feedback, emitter, 187
Degree of modulation, 287
Demodulation, 305, 308
Demodulators
amplitude, 314
series diode, 315
shunt diode, 316
transistor, 317
frequency, 318
Foster Seeley, 318
frequency discriminator, 318
ratio detector, 320
Depletion
area, 4
mode, 26
Detector, 308, 320
Deviation, frequency, 296
Devices, solid state, 1
unilateral, 133
Diaphragm, 282
Diffusion process, 12
Diodes
common, 1, 4, 5
photo, 18
power supply, 10
signal, 10
tunnel, 1, 4, 12
Zener, 1, 4, 15, 53, 58
Direct coupling, 158
Discriminators
Foster Seeley, 318
frequency, 318
Dissipation, power, 128
Distortion, signal, 126, 127, 197
Donor material, 3
Doping, 3

Double
 base diode, 26
 phase, 37
 tuned networks, 184
Drain, 23
Drivers, audio, 143, 161
Dynamic
 characteristics, 95
 current, 147
 microphone, 283
 speaker, 324
 transfer, 124
  applications, 126
  characteristics, 115
 voltage, 147

Eddy currents, 32
Effects of changing load, 122
Effective
 collector bias, 123
 Q, 180
Efficiency, transformer, 31
Electromagnetic waves, 281
Electron
 current and flow, 3
Electronic
 series regulator, 57, 58
 switch, 16
Elements of transistor, 19
Emitter
 current control, 189
 degeneration, 187
 injection,
  mixer, 311
  modulator, 290
Enhancement
 channel, 25
 mode, 26
Energy
 conversion circuits, 30
 losses,
  copper, 32
  eddy current, 32
  hysteresis, 32
Equivalent
 circuits, 96, 112
 parameters, 112
Establishing operating point, 118

Features of amplification, 65
Feedback
 internal, 132
 in tuned amplifiers, 185
 oscillator, 241
 resonant, 241
 windings, 60
Field effect transistors, 1, 22
 IGFET, 22
 JFET, 22, 23, 138
 MOSFET, 22, 25
Filters
 capacitor input, 44
 choke input, 46
 L sections, 47
 pi-type LC, 46
 power supply, 44
Forward
 bias, 6
 current, 10, 11
Free running oscillators, 60
Frequencies
 audio, 143
 power, 31
 signal, 196
Frequency
 center, 296
 cutoff, 73
 demodulators
  Foster Seeley discriminator, 318
  frequency discriminator, 318
 determining device, 239
 deviation, 296
 fundamental, 61, 197, 250
 harmonics, 197, 250
 limitation, 207
 modulation
  carrier wave, 295
  bandwidth, 299
  principles, 295
  sidebands, 298
 modulators
  capacitor microphone, 300
  reactance, 301
 problems, 131
 resonant, 175, 178
 response
  high, 207

ideal, 201
low, 202
practical, 202
RC, 201
ripple, 40, 42, 43, 46
stability in oscillators, 241
Full-wave rectifier, 41
Function generators
logarithmic, 223
servo techniques, 227
square law, 226
Functions
logarithmic, 218, 220, 223
nonlinear, 218
receiver, 305
square law, 218, 222, 226

Gain
common base amplifier
alpha, 71, 111, 137
power, 73, 138
voltage, 72, 138
common collector amplifier
current, 80
power, 83
voltage, 82
common emitter amplifier
ac beta, 76
beta, 75, 111
current, 77
power, 79
voltage, 77
control, 189, 192, 230
Gate, 23
Generators
function
logarithmic, 223
servo techniques, 227
square law, 226
sweep, 277

Half power point
high, 177
low, 177
Half-wave rectifier, 39
Harmonics, 197, 250
Heterodyning, 306

High frequency
amplifiers, 132
limitation, 207
response, 207
High level modulation, 291, 292
Hill, potential, 12
Holding current, 17
Hybrid parameters, 96, 102
designations, 102
for common base, 107
for common collector, 107
samples, 107
to open circuit, 109
to short circuit, 110
Hysteresis loss, 32

Impedance matching, 83
Increasing amplification, 206
Index, modulation, 299
Injection
base, 288, 309
collector, 291
emitter, 290, 311
Input resistance
common base, 72
common collector, 83
common emitter, 79
Insulator, 1
Interelement impedance, 131
Internal feedback, 132
Interrelationship of parameters, 109
Interstage coupling
direct, 158
impedance, 157
RC, 154
transformer, 156

Junction
diode, 218
field effect transistor, 23
avalanche breakdown, 140
load line, 139
N-channel, 138
operating point, 139
power curve, 139
temperature, 9

L sections, 47
Ladder oscillator, 254
Lag line oscillator, 254
Lamination, 32
Large signal amplifiers, 115
LC filter, 46
Load line, 95, 115, 136
   construction, 116
Loaded voltage divider, 51
Local oscillator, 307, 309
Logarithmic
   amplifiers, 229
      automatic gain control, 230
      constant output, 232
      dc amplification, 232
   functions, 218, 220, 223
   generators, 223
Low half power point, 177
Low level modulation, 290, 292
Low frequency
   attenuation, 203
   response, 202

Magnetic
   headphone, 322
   microphone, 284
   speaker, 323
Matching impedance, 35, 83, 180
Material
   acceptor, 3
   donor, 3
   N-type, 3
   P-type, 3
   pentavalent, 3
   trivalent, 3
Measuring parameters
   open circuit, 98
   short circuit, 100
Microphones, 281
   capacitor, 300
   carbon, 282
      pile, 282
   crystal, 283
   diaphragm, 282
   dynamic, 283
   magnetic, 284
Mixers, 309
   base injection, 311

   emitter injection, 311
   practical, 311
Mixing, 306
Modulating wave, 282
Modulation, 281
   amplitude, 285
      degree, 287
      high level, 291, 292
      ideal, 288
      low level, 290, 292
      percentage, 287, 299
      principles, 285
      sidebands, 286, 298
   frequency, 295
      bandwidth, 299
      carrier wave, 295
      center frequency, 296
      frequency deviation, 296
      index, 299
      principles, 295
      sidebands, 298
      significant sidebands, 298, 300
Modulators, 281
   amplitude, 288
      base injection, 288
      collector injection, 291
      emitter injection, 290
      oscillator, 292
   frequency, 300
      capacitor microphone, 300
      reactance, 301
MOSFET, 22, 25
Multiple loaded voltage divider, 51
Multipliers, voltage, 47
Multivibrators
   astable, 265
   bistable, 272
   monostable, 270

N-channel, 23, 25, 138
Negative
   carriers, 6
   rectifier, 42
   resistance, 12, 13
Networks, tuned, 184
Neutralization, 186
Noise factor, 149

Index   369

Nonlinear
  devices, 218
    junction diodes, 218
    optoelectronic, 222
    servomechanisms, 223
    thermistors, 222
    transistors, 220
    vacuum thermocouples, 222
    varistors, 221
    Zener diodes, 220
  functions, 218
    cube root, 222
    logarithmic, 218, 220, 223
    square law, 218, 222, 226
Nonresonant feedback oscillators, 253
  ladder, 254
  lag line, 254
  phase shift, 254
  Wien bridge, 255
Nonsinusoidal
  oscillators, 260
    relaxation, 260
    synchronized, 260, 269
  signals, 197
NPN transistors, 19
N-type material, 3

Open circuit parameters, 96, 97
  measuring, 98
  to hybrid, 110
Operating
  parameters, 66
  point, 84, 126, 139
Optoelectronic devices, 222
Oscillation principles, 237
Oscillators
  crystal controlled, 250
    Armstrong, 252
    Colpitts, 253
  nonresonant feedback, 253
    ladder, 254
    lag line, 254
    phase shift, 254
    Wien bridge, 255
  nonsinusoidal, 260
    blocking, 269

    multivibrators, 265, 270, 272
    relaxation, 260, 263, 264
    synchronized, 260, 269
  sinusoidal, 237
    amplifier, 238
    basic, 241
    common base Colpitts, 245
    common emitter Colpitts, 246
    feedback, 241
    frequency determining device, 239
    frequency stability, 241
    push-pull Hartley, 249
    series fed Hartley, 248
    shunt fed Hartley, 247
    tickler coil, 243
    tuned base Armstrong, 243
    tuned collector Armstrong, 244
Output resistance
  common base, 72
  common collector, 83
  common emitter, 75, 79

Parallel resonance, 174
Parameters, 96
  applications, 112
  converting, 107
  current gain, 112
  designations, 98, 100, 102
  equivalent circuits, 112
  hybrid, 96
  interrelationship, 109
  open circuit, 96, 97
  operating, 66
  power gain, 112
  short circuit, 96, 100
  variable quantities, 96
  variation of designators, 111
  voltage gain, 112
P-channel, 23, 25
Peaking
  series, 210
  shunt, 208
Peak inverse voltage, 11, 39
Percentage of modulation, 287, 299
Permanent magnet speaker, 323

Index

Phase
  double, 37
  shift oscillator, 254
  single, 37
  triple, 37
Piezoelectric effect, 250, 283
Pinch-off, 23
PNP transistor, 19
Positive
  carriers, 6
  rectifier, 42
Potential
  barrier, 7, 8
  hill, 12
Potentiometer, 50
Power
  amplifiers, 65, 67
    audio, 143, 164
  curve, 128
  dissipation, 128
  frequencies, 31
  gain
    common base, 73, 138
    common collector, 83
    common emitter, 79
  point
    high, 177
    low, 177
  supply
    diodes, 10
    filters, 44
    rectifiers, 38
  transformers, 30
Practical
  converter, 313
  mixer, 311
  wide band response, 201
Preamplifiers, 143, 149
  noise factor, 149
Predicting output, 119
Principles of
  amplitude modulation, 285
  amplification, 65
  frequency modulation, 295
  oscillation, 237
Process of diffusion, 12
P-type material, 3
Pulsating voltage, 40

Push-pull amplifier, 164
  low bias, 165
  no bias, 165

Quality of resonance, 176, 178
Quantum mechanics, 15
Quiescent state
  current, 144
  voltage, 144

Rarefaction wave, 281, 301
Ratings for
  power supply diodes, 11
  signal diodes, 10
Ratio detector, 320
RC
  coupling, 200
  frequency response, 201
  time constant, 260
Reactance frequency modulator, 301
Receiver functions, 305
Recovery time, 11
Rectifier
  bridge, 42
  circuits, 39
  full-wave, 41
  half-wave, 39
  negative and positive, 42
  power supply, 38
  silicon controlled, 1, 16
  three-phase, 43
Recurrency current, 11
Regenerative action, 61
Regulated voltage, 69
Regulator
  electronic series, 57, 58
  transistor shunt, 56
  voltage, 53
  Zener shunt, 53
Relationship of alpha and beta, 76
Relaxation oscillators, 260, 263, 264
Resistance
  input, 72, 79, 83
  output, 72, 79, 83
  surge, 41
Resistivity, 1, 7

Resonance
  conditions, 175
  parallel, 174
Resonant
  feedback oscillators, 241
  frequency, 175, 178
  quality, 176, 178
Reverse bias, 8
Rheostat, 50
Ringing, 212
Ripple frequency, 40, 42, 43, 46
Runaway, thermal, 9

Sample hybrid parameters, 107
Saturation, transistor, 86
Schmitt trigger, 276
Selectivity, 306
Series
  diode demodulator, 315
  peaking, 210
  regulator, 56
Servo
  mechanisms, 223
  techniques, 227
Shockley diode, 18
Short
  circuit parameters, 96, 100
  designators, 100
  measuring, 100
Shunt
  diode demodulator, 316
  peaking, 208
  regulator, 56
Sidebands
  AM, 286, 298
  FM, 298, 299
  significant, 298, 300
Signal
  diodes, 10
    ratings, 10, 11
  distortion, 197
  frequencies, 196
  phase relationship in
    common base, 73
    common collector, 82
    common emitter, 78
  waveforms, 145

Silicon controlled rectifier, 1, 6
Simple relaxation oscillator, 263
Single
  ended amplifier, 164
  loaded voltage source, 50
  stage phase inverter, 162
Sinusoidal
  oscillators, 237
  signals, 196
Small signal, 95, 115
  amplifiers, 95
  characteristics, 111
Source, 23
Space charge, 7
Speakers
  dynamic, 324
  permanent magnet, 323
Square law function, 218, 222, 226
  generator, 226
Static characteristics, 95
Stability factor, 88
Stabilization
  bias, 84
  circuits, 88
Steering diodes, 274
Summary of characteristics
  common base, 74
  common collector, 84
  common emitter, 79
Surge
  current, 41
  ratings, 11
  resistance, 41
Sweep generator, 277
Symmetrical Zener, 15, 54

Temperature, junction, 9
Thermal runaway, 9
Three-phase rectifier, 43
Tickler coil oscillator, 243
Time constants, 260
Tone control, 260
Transient current, 11, 39
Transformer
  bifilar, 198
  core lamination, 32
  coupling, 156, 181

372 Index

Transformer *(continued)*
  double phase, 37
  efficiency, 31
  energy losses, 32
  matching impedance, 35
  phase relation, 37
  power, 30
  ratios, 33
  single phase, 37
  triple phase, 31, 37
Transistor
  biasing
    input and output, 20
  bipolar, 1, 18, 65, 124
  current, 21
  cutoff, 85
  distortion, 87
  elements
    bipolar, 19
    field effect, 23
  field effect, 1, 22
  saturation, 86
  shunt regulator, 56
  unijunction, 1, 26
Treble control, 161
Trivalent material, 3
Tuned
  amplifiers, 174
    AGC, 192
      coupling, 174, 181, 182, 183
      feedback, 185
    base Armstrong oscillator, 243
    collector Armstrong oscillator, 244
    networks, 184
Tunnel
  diode, 1, 4, 12
  effect, 12
Tunneling, 12
Types of stabilizing circuits, 88

Unijunction transistor, 1, 26
Unilateral device, 133
Unilateralization, 186

Variable quantities, 96
Variations in parameter designators, 111

Varistors, 221
Voltage
  amplifier, 65, 67
  breakdown, 10
  control, 50, 191
  divider, 49
    bias, 69
  gain
    common base, 72, 78
    common collector, 82
    common emitter, 77
  multipliers, 47
    cascade doubler, 48
    doubler, 47
    tripler, 49
  pulsating, 40
  regulated, 69
  regulators, 53
    electronic series, 57, 58
    transistor shunt, 56
    Zener shunt, 53
  working, 10
Volume control, 159

Waves
  carrier, 281, 285, 295
  compression, 281, 301
  electromagnetic, 281
  modulating, 282
  rarefaction, 281, 301
  signal, 145, 196, 197
Wide band amplifiers, 196
  coupling, 198, 200
  response, 201
Working voltage, 10

Zener
  diode, 1, 4, 15, 53, 58
    characteristic breakdown, 54
    current, 54
    nonlinear, 220
    relaxation oscillator, 264
    shunt regulator, 53
    symmetrical, 16
    temperature compensation, 55
  region, 15, 54